PROVING GROUND

PROVING GROUND

The Untold Story of the Six Women Who Programmed the World's First Modern Computer

KATHY KLEIMAN

GRAND CENTRAL PUBLISHING

NEW YORK BOSTON

Cover design by Sarahmay Wilkinson
Cover images: top: (left) courtesy of the Bartik family; (center) by © Francis Miller/The LIFE Picture Collection/Getty images; (right) courtesy of the Hagley Museum and Library; bottom: (left) by © CORBIS/Getty Images; (center and background code image) Special Collections Research Center, Temple University Libraries, Philadelphia, PA; (right) University Archives and Records Center, University of Pennsylvania

Grand Central Publishing
Hachette Book Group
1290 Avenue of the Americas, New York, NY 10104
grandcentralpublishing.com
twitter.com/grandcentralpub

First Edition: July 2022

Grand Central Publishing is a division of Hachette Book Group, Inc. The Grand Central Publishing name and logo is a trademark of Hachette Book Group, Inc.

Library of Congress Cataloging-in-Publication Data

Names: Kleiman, Kathy, author.
Title: Proving ground : the untold story of the six women who programmed the world's first modern computer / Kathy Kleiman.
Description: First edition. | New York : Grand Central Publishing/Hachette Book Group, 2022. | Includes bibliographical references and index.
Identifiers: LCCN 2022004309 | ISBN 9781538718285 (hardcover) | ISBN 9781538718278 (ebook)
Subjects: LCSH: Women computer programmers—United States—Biography. | Computer programmers—United States—Biography. | ENIAC (Computer)
Classification: LCC QA76.2.A2 K56 2022 | DDC 004.092/2 [B]—dc23/eng/20220412
LC record available at https://lccn.loc.gov/2022004309

ISBN: 9781538718285 (hardcover), 9781538718278 (ebook)

Printed in the United States of America

LSC-C

Printing 1, 2022

To Betty, Kay, Jean, and Marlyn for sharing your stories so generously,
To Sam and Robin for listening to me tell and retell them, and
To Mark for joining in the journey

Contents

Cast of Characters

Electronic Numerical Integrator and Computer Programmers (the ENIAC 6)

Kathleen McNulty/"Kay"—Math graduate of all-women's Chestnut Hill College in Philadelphia. She was recruited to the Army's Philadelphia Computing Section at the Moore School in 1942 and became supervisor of a team operating the differential analyzer from 1942 to 1945. In 1945, she was chosen to program ENIAC, and moved with ENIAC to Aberdeen Proving Ground (APG) after World War II to continue her programming work.

Frances Bilas/"Fran"—Math graduate of Chestnut Hill College and Kay's best friend. She was recruited to the Army's Philadelphia Computing Section at the Moore School in 1942 and became the supervisor of a team operating the differential analyzer from 1942 to 1945. In 1945, she was chosen to program ENIAC, and moved with ENIAC to APG after WWII to continue her programming work.

Frances Elizabeth Snyder/"Betty"—Graduate of University of Pennsylvania. She joined the Army's Philadelphia Computing Section at the Moore School in 1942. In 1945, she was chosen to program ENIAC, and moved with ENIAC after WWII to APG to continue her programming work. She became an early employee of Eckert-Mauchly Computer Corporation working with the first commercial computers and went on to a forty-year career at the cutting-edge of computing and programming.

Marlyn Wescoff—Graduate of Temple University. She joined an Army radar project at the Moore School in 1942 and became part of the Army's Philadelphia Computing Section in 1943. In 1945, she was chosen to program ENIAC's ballistics trajectory program.

Ruth Lichterman—Undergraduate studying math at Hunter College in New York City when she was recruited to the Army's Philadelphia Computing Section at the Moore School in 1943. In 1945, she was chosen to program ENIAC and moved with ENIAC after WWII to APG to continue her programming work. She then returned to the Moore School to work on other projects.

Jean Jennings—Math graduate of Northwest Missouri State Teachers College in Maryville, Missouri. She moved to Philadelphia in 1945 to join the Army's Philadelphia Computing Section. In 1945, she was chosen to program ENIAC. After WWII when ENIAC moved to APG, she continued her work with ENIAC, helping to design and program its Converter Code and hiring and leading a programming team to deliver the Ballistic Research Laboratory's (BRL) first wind tunnel programs for ENIAC. She continued at Eckert-Mauchly Computer Corporation and had an active career in computing and computer publishing.

Moore School of Electrical Engineering, University of Pennsylvania

Dr. Harold Pender—First dean of Moore School. He contracted with the BRL for Moore School to host the Army's Philadelphia

Computing Section during WWII. Under his tenure, J. Presper Eckert and Dr. John Mauchly built ENIAC.

Dr. John Grist Brainerd/"Grist"—Moore School instructor, dean, and Director of Research. He was the liaison to BRL at APG for the Army's Philadelphia Computing Section and later the ENIAC project. He was close to Lieutenant Herman Goldstine.

Joe Chapline—Beginning in May 1942, he was a research associate at Moore School and a maintenance engineer on the differential analyzer. He was a champion of John's computing ideas and connected Herman Goldstine with John Mauchly.

Dr. John Mauchly—Received his PhD in physics from Johns Hopkins University. He was a professor at Ursinus College when he left to answer the Army's call for men with electronics abilities during World War II. He was cofounder of Eckert-Mauchly Computer Corporation and co-inventor of the world's first general-purpose, programmable, all-electronic computer, ENIAC, and its successors BINAC (Binary Automatic Computer) and UNIVAC (Universal Automatic Computer), the first modern commercial computers.

J. Presper Eckert Jr./"Pres"—Electrical engineering graduate of the Moore School. He was a lab instructor there when Mauchly first met him. He was cofounder of Eckert-Mauchly Computer Corporation and co-inventor of ENIAC, BINAC, and UNIVAC and was considered one of the finest electrical engineers of the twentieth century.

Dr. Irven Travis—Electrical engineering professor at the Moore School who was called to active duty in 1941. He returned in 1946 and became Director of Research. He ended up clashing with John and Pres over patent rights.

Ballistic Research Laboratory, Aberdeen Proving Ground, Aberdeen, Maryland

Colonel (later Major General) Leslie E. Simon—Director of BRL during WWII. He created the Army's Philadelphia Computing Section at the Moore School during WWII with Captain Paul Gillon and worked with John Grist Brainerd and Herman Goldstine. He negotiated and approved the contract with the Moore School to build ENIAC and relocate it to BRL after its acceptance by the Army.

Captain (later Colonel) Paul N. Gillon—Assistant Director of BRL during WWII. He was Herman Goldstine's senior officer and cocreator of the Army's Philadelphia Computing Section at the Moore School with Colonel Simon. He supported BRL's funding of ENIAC and worked with the Moore School team from time to time.

Lieutenant (later Captain) Herman Goldstine—PhD in math from University of Chicago, where he studied under Dr. Gilbert Bliss. After being sent to BRL during WWII, he was assigned to head the Army's Philadelphia Computing Section at the Moore School and served as liaison from BRL to the Moore School. He introduced the BRL to the idea of ENIAC and was involved in its construction and demonstration. After WWII, he was a member of Princeton's Institute for Advanced Study (IAS) and then went to IBM.

Adele Goldstine (née Katz)—Graduate of Hunter College, with a master's degree in math from University of Michigan. She revamped instruction at the Army's Philadelphia Computing Section at the Moore School to educate dozens of young women in graduate-level numerical analysis for trajectory calculations.

She wrote the technical manual for ENIAC and after WWII helped design the ENIAC Converter Code and helped convert ENIAC to one of the world's first stored-program computers. She was the wife of Herman Goldstine.

John Holberton—Beloved civilian supervisor of the Army's Philadelphia Computing Section, he reported to Lieutenant Goldstine and worked with him on the ENIAC project too. He moved with the ENIAC to APG after WWII and spent a career in computing.

Major General Oswald Veblen—Professor at Princeton for decades and early member of the Institute for Advanced Study. He joined APG during WWI and worked on ballistics research and calculations. He founded a program to calculate ballistics trajectories for APG, which he insisted continue between WWI and WWII. He was named Chief Scientist of BRL during WWII and made the final decision to fund ENIAC.

Preface

As I stared at the women in the black-and-white photograph, it seemed as though they were trying to tell me something. I was sitting in Harvard's Lamont Library, a main library for undergraduates, trying to research a paper on American women in the twentieth century who were leaders in computing. I knew of only one, Captain Grace Hopper of the US Navy, later Rear Admiral Hopper. Of course, there was Lady Ada Lovelace, daughter of British poet Lord Byron, who worked on early programming concepts in the nineteenth century, but she was out of scope for an American women's history course.

I was a young woman in computing, and I wanted to know if there were others. I had taken computer science since I started college, and while my early programming courses were composed of about half women, my latest class had only one or two. I knew I would feel more comfortable in computing if I saw a few more women in class with me, and this drove my interest in who came before me and what they had done.

Open before me, spread across the reading room table, were encyclopedias of computer science and histories of computing. Noticeably absent in all of them were the names of women, except Ada and Grace. But noticeably absent, too, was any real history of programming. The stories were all about hardware and the men who built the mainframe computers that dominated computer history in the 1940s, '50s, and '60s.

But what about those who pioneered ways to communicate with the large computers? Instruction codes and programming languages also date back to the 1940s, '50s, and '60s, but where were the stories of the people who wrote them?

Then I stumbled on a black-and-white photograph of a huge, black, metal computer dominating three sides of a large room and dwarfing six people—four men and two women.

University Archives and Records Center, University of Pennsylvania

There were two men in the middle of the photo, two women on the right with a man in uniform between them, and a man in the back left. Only the two men in the middle were named—J. Presper Eckert and Dr. John Mauchly, co-inventors of ENIAC, the world's first all-electronic, programmable, general-purpose computer. It was built at the University of Pennsylvania (Penn) during World War II. Nowhere in the captions or accompanying article were the other people in the photograph named.

I studied the image closely, especially the women. They were young, with World War II–era hairdos, flat shoes, and skirt suits. As I leaned in closer, what struck me was that they seemed to know something about ENIAC. They appeared to be comfortable, knowledgeable as they adjusted knobs on and read documents next to this vast, seemingly living and breathing giant. I could not stop looking at them.

I knew something about computers. My father was an electrical engineer who specialized in new technologies. He brought home electronics for us, including an early calculator, clunky and huge compared to today's versions, with only a few functions, but fascinating and fun to play with. He was the first person I knew to talk about speech synthesis and voice recognition. My father had written his dissertation on the founding of the semiconductor industry and was certain that the miniaturization of electronics would continue and would keep changing the world.

Friends asked me in junior high school if I wanted to learn to program computers, and I said yes. So I joined an Explorer Post, a coed branch of the Boy Scouts dedicated to career exploration, and went off to spend my Wednesday nights at Western Electric, a manufacturing arm of AT&T near my home in Columbus, Ohio. I learned BASIC, a programming language invented at Dartmouth in the 1960s. Soon I was playing games my friends wrote and adding one of my own, a version of Mad Libs that I wrote in BASIC. The first time my friends played it and laughed out loud at the funny story the computer printed, I knew I was hooked on programming.

Looking at the old black-and-white picture, and the men and women standing before ENIAC, I longed to know more about their story. I dug deeper, found more books, and uncovered another photograph. This one was a close-up showing two young women standing right in front of ENIAC. Once again, no names of the women, only the name of the computer.

University Archives and Records Center, University of Pennsylvania

I made copies of both pictures and took them to my professor. Anthony Oettinger was a former president of the Association for Computing Machinery, an international group for computing professionals.

I showed him the two ENIAC photographs. "Who are the women?" I asked.

"I don't know," Professor Oettinger answered, "but I know who might."

He told me to visit Dr. Gwen Bell, cofounder of the Computer Museum, then in Boston and now in Silicon Valley.

The Computer Museum was at the far end of Museum Wharf in downtown Boston. As I walked down the long wharf, I noticed the Children's Museum and in the water, the Boston Tea Party ships. All great to visit, but I was on a different mission as I clutched my folder of photographs and disappeared into the Computer Museum.

I found my way to the office of the museum's director. Bell was in her early fifties with short, dark, graying hair, clearly a no-nonsense, busy person. I opened my folder and once again found myself pointing to the black-and-white images of the women who were standing in front of ENIAC.

"Who are the women?" I asked Bell.

Unlike Professor Oettinger, she knew. "They're refrigerator ladies," she said.

"What's a refrigerator lady?" I asked, baffled as to what she was talking about.

"They're models," she responded, rolling her eyes. Like the Frigidaire models of the 1950s, who opened the doors of the new refrigerators with a flourish in black-and-white TV commercials, these women were just posed in front of ENIAC to make it look good. At least that's what Dr. Bell thought.

She closed my folder and handed it back to me. I was dismissed.

I slowly headed out of the museum. I saw the children lined up on the wharf getting ready to enter the Children's Museum, the Boston Tea Party ships rocking in the harbor, and the bright blue sky. But Bell's story did not make sense to me. I had stood in front of big computers before. The first time you see them, they seem huge, overwhelming, almost surreal. The young women in the ENIAC pictures looked confident and assured. They looked like they knew exactly what this huge computer did and why they were in the photos. They *did* appear posed, but so did the men, and the men were not models.

As I left the wharf, I set a task for myself. I was going to find out the names of the women. I would learn what they had done in order to be in these beautiful, black-and-white pictures of ENIAC in the 1940s.

I was going to learn their story.

PROVING GROUND

The Double Doors Open

The women walked down the stairs from the second floor, their saddle shoes squishing against the marble. All of the students and professors at the Moore School of Electrical Engineering were men, and the women were used to being hazed, ogled, and mocked when they walked down the hallways or down to the differential analyzer room, where some of them had worked.

But this time the hallways were quiet. Betty, Jean, Kay, Fran, Marlyn, and Ruth—the six women assigned to program solutions for Army ballistics trajectory problems—all twentysomethings, turned right and stopped at a set of double doors. A sign read RESTRICTED. For three and a half months, in classrooms, an antechamber, and a repurposed nearby fraternity house, they had been familiarizing themselves with the computer behind the door, studying diagrams and piecing together the puzzle of how to use it. But they had never once been granted the privilege of seeing the computer that they would program. They had been prohibited from setting eyes on it—until now.

It was mid-November 1945, almost three months after the official surrender of the Japanese, and the women's boss, Captain Herman Goldstine, serious-minded with an officious air, was walking in front of them. With no warning, he had summoned them down from the second-floor classroom in which they had been working.

The double doors swung open, and they came face-to-face with the Electronic Numerical Integrator and Computer—the great ENIAC, all forty-five units of it. They had nearly given up asking when they would see it, let alone move the switches and wires and cables they had come to know so intimately on paper over the past three months. But now everything had changed, and they came out of the classroom and followed Herman down the wide concrete steps to the first floor.

It was like meeting someone in person for the first time, after studying their face in photographs. A series of black steel units, each eight feet tall and two feet wide, stared down at them. The units stood in the shape of a huge U—sixteen on the left side, eight at the base, sixteen more on the right. Three square units on wheels stood at odd places around the room, while the remaining two units, an IBM card reader and an IBM card punch, were connected by wires. The women were delighted.

Curious to get a better look, they walked around the large thirty-by-sixty-foot room. The units had been pushed away from the wall so that the engineers could work on the backs. The women took in ENIAC's depth, a few feet. They inspected the great U and examined the units and their switches. They were impressed: They knew this huge computer was capable of completing 5,000 additions in a second and 500 multiplications in the same second, not to mention lightning-fast divisions and square roots.

Even with the AC roaring, they could feel the heat coming off the tubes and hear their low hum. It had taken more than 200,000 hours of work to build ENIAC and cost just under $500,000—equivalent to just over $7 million today. The women had closely studied the machine on paper, but to see it before them in real life was surreal. They walked around the room completely absorbed by ENIAC, oblivious to the other people in the room.

Too soon, their senior supervisor Herman Goldstine brought them out of their reverie with his command: "We're going to put a problem on." They looked up and realized there were about a dozen people already inside the room that they had missed while gazing around in wonderment. These included some young ENIAC engineers (builders of the computer) and Herman's wife, Adele, a mathematician who had trained some of the women to calculate ballistics trajectories when they first started their Army work. There were also two men from an Army base in New Mexico who the women had met briefly that summer, Dr. Stanley Frankel and Dr. Nicholas Metropolis.

Herman quickly assigned the women to work with various other people in the room, and they took their places around the units of ENIAC. Metropolis and Frankel distributed small, prepared slips of paper that were slid into metal slots at the front of many of the units.

Engineers had been testing small problems on ENIAC, such as tables of squares and cubes, but this new problem, which no one explained, seemed to take up almost the whole machine. As everyone awaited the next command, there was a moment of silence.

Standing in the center of the room, Herman lifted his hands like an orchestra conductor. It would direct the actions of the entire group as they strung wires across the computer and hoisted 50-pound digit trays into place.

The young women were about to do something they had never done before, but they had a shared history that made them exuberant and optimistic. They would work on ENIAC the same way they had done everything since they began their collective journey more than three years earlier: by sitting at desks typing numbers into clunky calculators; by holding what they liked to call "bull sessions" in their barracks; by squinting at enormous diagrams to learn ENIAC's units in borrowed rooms on the Penn campus. They would teach themselves to program a computer they were not even allowed to see.

And they would do it together.

Looking for Women Math Majors

On a cloudy day on Tuesday, June 2, 1942, Kathleen "Kay" McNulty, twenty-one years old, smiled and took her bachelor's in mathematics diploma from Reverend Hugh L. Lamb, Auxiliary Bishop of the Archdiocese of Philadelphia.[1] She had dancing eyes, a narrow face, and dimples. It was commencement at Chestnut Hill College, a Catholic women's school in the northwestern edge of Philadelphia, overlooking the Wissahickon Creek. Kay was one of 107 graduates.[2]

The commencement was outside, near the tennis courts, and the principal address was also delivered by Bishop James Kearney of Rochester. One of Kay's best friends, Frances "Fran" Bilas (pronounced *BEE-las*) received many awards that day, including one from the National Catholic School Press Association, the Student Teacher's Gold Key Award, and a Kappa Gamma Pi certificate "for graduation with distinction and leadership in extra-curricular activities."[3] Kay knew Fran was one of the smartest students in the class. As Kay and Fran met their families, clutching their degrees in their hands, both of them knew that they were starting the next chapters of their lives.

It was a strange time to be a young American entering the job market. At the University of Pennsylvania commencement exercises, which took place the same day, seventy-three degrees were awarded in absentia to young men who had already joined the armed services.

The *Philadelphia Inquirer* might have been addressing Fran and Kay with the headline it ran above photos of different area commencements: STUDENTS GRADUATING INTO WORLD AT WAR.[4]

The *Women's Undergraduate Record*, the yearbook for women at the University of Pennsylvania for 1941, declared that the war was

> a great sorrow to the world... Even though we were not actively engaged in it, it is a war world in which we live. We cannot isolate our sympathies, even though we may hope to isolate our nation. Our eyes are on Europe and her guns strike our hearts. The maturity of Seniority has been accepted by the sober thoughts that are with us all.[5]

Kay and Fran had been two of only three math majors in their class at Chestnut Hill College; the third, Josephine Benson, was also their best friend. Kay had picked math because it was easy and fun for her. A few days into college, an adviser asked her to pick a major, telling her to choose the subject she liked best. "Mathematics," she immediately responded. For her, math was "no work. It was a no brain thing for me. It was just like a wonderful puzzle that you could do and there was always an answer."[6]

Most women who entered Chestnut Hill College during the Depression majored in home economics, the study of life skills such as cooking, sewing, and finance. In fact, just a few weeks before Kay, Fran, and Josephine's commencement, Chestnut Hill's home economics department presented a fashion show in the school auditorium. A hundred gowns were modeled by students, with a patriotic theme, due to the war.[7] Many young women in Kay's class wanted to be dietitians in schools or hospitals. Then they would marry and have children. Home economics could help them in the dietitian field, but that wasn't the point; they had to learn how to cook and run a household well if they were going to be good housewives.

Kay was not like many of the other students. She wanted to do something important, and eventually she wanted to start a family. And she did not think the two were mutually exclusive.

Not two weeks after she graduated, she spotted a notice in Philadelphia's *Evening Bulletin*: LOOKING FOR WOMEN MATH MAJORS. The Army sought women to work at the University of Pennsylvania's Moore School of Electrical Engineering. She didn't know what the job was but thought it amazing that a job for *women* with degrees in mathematics "would be advertised in the paper."[8] Before the war, this would have been unheard of; ads for math-related positions (such as accountants and actuaries) were in the "Male Help Wanted" section of the newspaper. Math was a man's job. In the "Female Help Wanted" sections, there were jobs for secretaries, nutritionists, nannies, and laundresses.[9] Those interested in the Moore School opportunity were to report to a recruiting office on South Broad Street in South Philadelphia, inside the Union League, a storied private club that also contained offices.[10] Kay called Fran and Josephine and said they should all interview together.

But Josephine already had a job. And so the next day, Kay showed up with only one best friend, not two.

All over the country, American women were seeing notices telling them they were needed for war work. Many of these ads were for industrial positions. With brothers, cousins, uncles, and fathers volunteering for service and being drafted, the government and military began a deliberate strategy to recruit women into factories and farms, for now-vacant positions.

During the Great Depression, it had been difficult for both men *and* women to obtain jobs. Unemployment soared to 24.9 percent in

1933 and remained above 14 percent from 1931 to 1940.[11] During World War II, the government encouraged women to fill the jobs formerly open only to men—and women enthusiastically responded. From 1940 to 1945, the percentage of women in the workforce increased by 50 percent.[12]

If the country were to clothe, feed, and provide guns, artillery, planes, and tanks to the armed forces, its women would have to take jobs in industrial manufacturing and in labor. The fictitious Rosie the Riveter, later the subject of a WE CAN DO IT! poster, was first introduced in a 1942 song that went, "All day long whether rain or shine / She's a part of the assembly line / She's making history, / working for victory." Rosie had "a boyfriend, Charlie / Charlie he's a Marine / Rosie is protecting Charlie / Working overtime on the riveting machine."[13] Posters recruiting women to war work trumpeted, "The more WOMEN at work the sooner we WIN!"[14]

Millions of American women stepped up to the plate—taking on jobs making jeeps and tanks, sewing uniforms, canning food, making weapons and ammunition, and doing production for the wartime movies that kept the population (at home and overseas) entertained and diverted. Norman Rockwell's Rosie the Riveter picture, featuring a young woman in blue overalls holding a sandwich in one hand with a rivet gun on her lap, and published on the cover of the *Saturday Evening Post* in May 1943, "proved hugely popular" and the *Post* loaned it to the US Treasury Department for war bond drives for the rest of the war.[15]

Economic mobilization due to the war shifted many boundaries of traditional "men's" and "women's" work. Previously male-defined jobs such as building tanks and repairing airplanes were recast as feminine and glamorous, and women were welcomed—for the time being.[16]

But separate from the boom in industrial employment, the war greatly expanded opportunities for college-educated women with backgrounds in engineering, science, and math. Women like Kay were seeing notices that seemed to be written just for them.

The Department of Labor's Women's Bureau proclaimed the new opportunities:

> [For] war job opportunities in science and engineering, you will find that the slogan there as elsewhere is "WOMEN WANTED!"[17]

Women with math degrees were desirable assets who could help the Allies win the war.

Several months after Kay saw the ad in the *Evening Bulletin*, leaders of war industries and women's college heads met at the Washington, DC, branch of the American Association of University Women to discuss steps that could speed the induction of college-trained women into specialized war jobs.[18] During the conference, vice chairman of the War Production Board William Batt told the *Philadelphia Inquirer*, "Winning this war is a job of great magnitude and tremendous seriousness, and there is an appalling demand of everything an Army needs." Women, he said, were demonstrating that "they can do as good a job as men, and in many instances a much better job in the factories and shops." He said American women had not yet reached the pace of women in Russia and England in their war activity, but they could.[19]

For decades, college-educated and highly skilled women had been turned away from well-paying jobs despite their qualifications. Now they held no grudges and jumped into the workforce wherever they were needed.

When Kay and Fran went to the office inside the Union League, they were met by an Army recruiter who asked them about their math backgrounds right away: "Did you have differential calculus?"

"Yes," Kay answered.

"Did you have physics?"

"Yes."

"You are exactly what we need," he said.

"They hired us on the spot," Kay remembered.[20]

She was happy; she had been out of college for only two weeks and she had a job—somewhat unusual for a graduate in any era, but even more unlikely given the circumstances. She and Fran were to report to the Moore School on July 1, 1942.[21]

Kay was born in Creeslough, whose name means "surrounded by lakes." It is in the northwest part of Ireland, in Donegal County, and Kay was born on the same land where her father's family had lived since 1804. Their land was 160 acres and ran from the top of Cruckatee Hill and its lake down to the ocean.[22]

Her father, James McNulty, was the youngest of seven children, and both of his parents had died while he was a child. An older brother had gone to the United States and was studying to be a stonemason, and James had three uncles who had also emigrated. He decided to take a three-year apprenticeship in Philadelphia to become a stonemason as well.

While serving his apprenticeship, James was active in Irish politics in Philadelphia, becoming an Irish Volunteer and studying how to drill and train troops. The group raised money for guns and ammunition and trained to return to Ireland to throw out the British. He was also a champion Irish step dancer and won many medals for it. In 1915, he got typhoid fever and went home to Ireland to recover.[23]

James and Annie married in February 1917 on the McNulty family farm. In 1918, Kay's older brother Patrick was born. A year later, her brother James (Jim) was born, and two years later, on February 12, 1921, Kay arrived.

The night Kay was born, her father was arrested. He had been with a group of men who blew up a bridge and were in hiding. Knowing the baby's birth was imminent, James returned home to be with Annie for the birth. "Name her Kathleen after my mother and grandmother," he said, and then was arrested by the Black and Tans, English recruits to the Royal Irish Constabulary during the Irish War of Independence. Many Irishmen were arrested the same night, but most of the men who had been with James at the bridge were not caught. For two years he was kept in Derry Jail in solitary confinement without any charges ever being brought against him and no trial. When he was brought to trial, without any charges, he was released.[24]

After his release from jail in 1923, James tried to live in Ireland but decided he could not live under the new Irish Free State, established in December 1922 under the Anglo-Irish Treaty. He opted instead to return to America and have his family join him. He and a brother formed a development company to buy land and build houses in Chestnut Hill, Philadelphia, a beautiful, tree-lined, growing suburb of Northwest Philadelphia.

While James was building a house for his own family, Annie, who was pregnant, delivered Kay's sister Anna in Ireland. In October 1924, Annie and the children sailed for the United States. James met them in New York and took them to their new, furnished home in the Wyndmoor section of Philadelphia.[25]

Kay arrived at the age of three speaking only Gaelic. But she learned English quickly, from books her brothers brought home from school. As her English grew stronger, she dropped her Irish accent, and yet whenever she came in from outside and stepped across the doorstep of her home, it came roaring back.[26]

After another baby, Cecelia, was born, the McNultys moved to a larger house, on Highland Avenue in Chestnut Hill. When Kay entered the local, almost entirely Irish, Catholic elementary school at six and a half, she was adept at math and advanced for her grade. One day her teacher announced that she was going to teach the class to count to ten. Kay stood up and said she could already count to fifty. "You don't need to be here," the teacher said. "You can go home." Kay walked out of the school.[27]

The teacher ran after her down the block, laughing. "I was joking!" she exclaimed. "You can stay! I'm going to teach you other things besides how to count." In the second half of third grade, Kay was advanced to fourth grade. She loved everything about school, and on the way home would stop at the local library to check out books. Sometimes she would sit on the steps in front of a bank and tell stories to other children, stories drawn from books and from her imagination. It got to be so late that her mother Annie would send one of the boys to make her come home.[28]

At night, during homework time, Kay helped her older brothers with their math homework.[29] As a result, she learned each math course at least a year before she studied it.

Other people were starting to notice her math skills too. At a nearby neighborhood store, the owners, two sisters, totaled up purchases with a pencil on the back of a paper bag. Kay could add up all the prices in her head faster than it took the sisters to write, and that impressed them. "When you grow up, you should be a mathematics teacher," one of the sisters told her.[30] Kay remembered the compliment for years to come.

Throughout her childhood, Kay was intrigued by gadgets. When her mother pulled the plug out of the iron while ironing, she gave Kay a screwdriver and let her fix it. When she was sixteen, she learned to drive, just as her older brothers had done. It didn't matter that she was the only girl in her neighborhood with a driver's license.[31]

After the stock market crash in 1929, the housing development market dried up, and Kay's father, James, went to work for Jack Kelly, a famous builder in Philadelphia and Washington, DC. In the late 1930s and 1940s, James did a lot of work in Washington as part of the teams that built the Jefferson Memorial and the Pentagon. He commuted home every other weekend.[32] Jack's daughter, Grace Kelly, was a few years younger than Kay and became a film star and later Princess of Monaco. The two men were friends for the rest of their lives.

The McNultys had an active social life and both loved to dance. They attended Catholic Church dances and brought the family to parties at the Kelly home. Whenever James came back for the weekend, he would plan an outing, like canoeing on the Wissahickon in Fairmount Park. On John Barry Day each year, which honored the Revolutionary-era Irish American Continental Navy officer and "father of the American Navy" commodore, the family went to the celebration by Independence Hall.[33]

Annie taught the girls Irish crocheting, Irish embroidery, knitting, table setting, and table manners. Kay learned to bake pies, and from her mother's sewing work became interested in making her own clothes.[34]

When she graduated from grammar school, she received a prize for perfect attendance. She also received first prizes for her grades and for her handwriting. She entered John W. Hallahan Catholic Girls' High School, which had 4,000 students and was an hour and a quarter away. She commuted by trolley, subway, and on foot. She worked on the newspaper and was on the honor roll for three years, only missing one year when Annie grew ill and she had to care for her.[35] She studied math, Latin, French, science, biology, chemistry, and every math course the school offered: algebra, plane geometry, advanced algebra, solid geometry, and trigonometry.[36]

She graduated from Hallahan in 1938 and watched as many of her fellow students went directly into the labor force, working for low salaries in department stores or in secretarial roles, taking dictation and typing. Those who attended college often chose two-year colleges called "normal schools" to become teachers. But Kay wanted the four-year path, and when Chestnut Hill College, a few miles away from her house, offered her a merit scholarship, she took it.

Chestnut Hill College, with its beautiful stone buildings with tall red turrets, stands atop a hill on the west side of Philadelphia. It opened in 1924 as a Catholic, four-year, liberal arts college for women. Originally called Mount Saint Joseph College and run by the Sisters of Saint Joseph, it was renamed in 1938, the year Kay entered.

Kay took what was then called "college algebra" (now precalculus), continuing with spherical trigonometry, integral calculus, differential calculus, and differential equations. She also studied several kinds of geometry, astronomy, and two years of physics.[37]

As much of an academic star as Kay, Frances Veronika Bilas was born in Philadelphia on March 2, 1922, the second of five girls. Her father, Joseph, was born in Yugoslavia and was a district engineer for the Philadelphia Board of Education, responsible for fifty-two school buildings. Her mother, the former Anna Hughes, was an elementary school teacher who returned to teaching after the girls grew up.[38]

Fran graduated from South Philadelphia High School for Girls in January 1938, at age sixteen. Like Kay, she also had a scholarship to Chestnut Hill College.[39] Her commute was an hour and a half each way. Intense, smart, and hardworking, Fran had hair so long it reached below her waist and she wore it braided up on her head. At Chestnut Hill, Fran, who was shy and self-conscious, majored in mathematics and minored in physics.[40] She was solemn, bookish, and kept to herself. Though she and Kay were close, she was a one-on-one person, not a mingler by nature.

Fran took an internship teaching at Simon Gratz High School, with the intention of becoming a math teacher.[41] Built in 1925, Gratz had been named in honor of the civic leader Simon Gratz, a significant innovator in twentieth-century Philadelphia schooling.

In math club, Fran, Kay, and Josephine grew close, spending much of their time in the "day hop lounge," which was available to the many students who commuted to the college and wanted to spend time working and chatting together. Kay also worked in the college bookstore and came to know almost everyone on campus. She was popular and well-liked there, and always eager to help fellow students find what they needed on the shelves.[42]

Approaching the end of college, Kay decided to add a minor in business administration to her major in mathematics, taking courses in accounting, money, and banking. She wanted to land an office job in insurance or business.[43] In the 1940 census, women made up 54 percent of all clerical workers; the number would rise to 62 percent in 1950.[44] In the summer of 1941, before Kay's senior year, as the prospect of US involvement in an international war loomed, she went to an employment agency to get a summer job. The recruiter told her she had an opening for a bookkeeper. Kay said she could do it, though in fact her accounting courses had not covered it extensively. That day was a Friday and the job started on Monday morning. She went to the library and checked out every single book on bookkeeping, spending the weekend learning everything she could. On Monday morning, she showed up.[45] Evidently her cramming had worked because she was hired and retained for the entire summer.

Although the United States had not officially entered World War II, by 1941 the Army had been recruiting heavily and Kay's brothers Pat and Jim enlisted in the Navy. Pat had been a budding inventor but wanted to do his part for the defense of his country. Kay's life, and the country's life, changed on the morning of December 7, 1941, when the Japanese Navy committed a surprise strike against Naval Station Pearl Harbor in Hawaii. The base was attacked by 353 Japanese fighters, torpedo planes, and bombers. All eight Navy battleships were damaged, and four of them sunk. The McNultys knew boys from Chestnut Hill who died at Pearl Harbor, promising young men whose whole lives had been in front of them.[46]

The attack was a shock to all Americans. In a radio address heard by Kay and millions of others the following day, President Franklin D. Roosevelt marked December 7 as a "day that will live in infamy" and announced the entrance of the United States into the war. From the start, the country would be fighting a war on two fronts: the Pacific and Europe.

Philadelphia transformed almost overnight. Before the war, there was high unemployment and empty factories—due to the Great Depression. But in June 1940, after France was taken by German armies, the Roosevelt administration began a rearmament program, boosting Philadelphia's economy, as government orders poured in for supplies and arms. Greater Philadelphia became known as an "arsenal for democracy"; it already possessed a Navy yard, arsenal, and plentiful universities.[47] Men and women streamed in from the North, South, and West.

By the time of the Pearl Harbor attack, industry in Philadelphia was revived. The Delaware River ran through the city, connecting it to the Atlantic Ocean. The Philadelphia Naval Shipyard, built in 1801, would grow from a few thousand workers in 1939 to 58,000 at its peak. By 1944, shipyards nearby in Kensington, Camden, and Chester would employ more than 150,000 workers, with the Delaware River–built ships a crucial contribution.[48]

Work to support the wartime effort was the first priority for everyone. The second was to roll out the red carpet and host dances for the soldiers and sailors in area camps and installations. Many of these dances took place at the Stage Door Canteen in the basement of the Academy of Music, a big dance floor near City Hall. Buses brought young soldiers from the bases where they were training to meet local young women on the dance floor and do both the lively swing and slow, romantic foxtrots.[49]

When they were not working long hours or entertaining soldiers, people listened intently to the news from their homes. Families gathered around their radios to hear Edward R. Murrow's crackly broadcasts from London on CBS News.[50] In 1940, they learned of the bombings of British citizens as the German Air Force hit London, night after night, in a blitzkrieg, with the British Royal Navy fighting to push them back. Hearing the earsplitting explosions and German Luftwaffe planes flying low and close to the British population helped galvanize the United States to aid their close cultural neighbors overseas.

It was only six months after the "live in infamy" address that Kay and Fran went to the Union League and were hired by the Army's Ballistic Research Laboratory (BRL), based at Aberdeen Proving Ground (APG) in Aberdeen, Maryland. Their role would be with the Army's Philadelphia Computing Section at the Moore School, and they would share the same title: Assistant Computer.

The term *Computer* had been used as far back as the seventeenth century to describe men who tracked time in calendars. During the late 1860s, when the astronomer Maria Mitchell was working for the US Coast and Geodetic Survey, it meant "a person who computes."[51] For years it described people, not machines.

Fran's and Kay's civil service salary was $1,620 per year, or about $27,000 today. Their civil service grade was "SP-4."[52] *SP* stood for "subprofessional and subscientific," a classification that had existed since the 1920s for duties "subordinate or preparatory" to the work of employees in the professional and scientific service.[53] It was a misnomer because these positions were clearly professional, calling for mathematical and science backgrounds. But the Army knew that the young women were so anxious to help their country they would not argue over rank. The rule was simple: Only men were allowed to get the professional "P" ratings.[54] Ultimately, even an astronomer with her PhD from Harvard would be classified as SP.[55]

As Kay walked out of the Union League on that hazy June day in 1942, she was one of those women who would not be arguing over her rank. In fact, she was delighted with her new job. She would be helping an Army project using the best of her abilities—her math skills. Plus, she would make more than double the going salary for a secretary, and she did not know anyone making that much. Because her job was classified, there was one key question she would not be able to answer until she reported to the Moore School for her first day: What on earth did an Assistant Computer *do*?

We Were Strangers There

The two graduates reported for their first day on July 1, 1942. They walked up the main entrance of the Moore School Building on South 33rd Street into a vestibule, and then went up six more stairs into a lobby before heading for the second-floor classroom. The University of Pennsylvania had launched an engineering program in the mid-1850s, but it was not until 1893 that the Department of Mechanical and Electrical Engineering was established. In 1923, after receiving a bequest from Alfred Moore, whose family made a fortune making high-quality wires for telegraphs, telephones, and electricity, Penn renamed it the Moore School of Electrical Engineering.

With Moore's money, Penn purchased the Howard E. Pepper Building at 33rd and Walnut Street, formerly manufacturing horns and publishing music, and converted it to classrooms, labs, and offices. In 1926, the Moore School moved in and the school grew to meet, among other needs, Philadelphia's growing demand for engineers to work in generating and distributing electrical power. It also upheld one particular requirement in Moore's will. He wanted only male students at his new school, no female students and faculty, and until the 1950s, there were none.[1]

Although the Moore School was considered excellent, former *Philadelphia Inquirer* science editor Joel Shurkin described it as living "deep in the shadow of MIT," respected but lacking prestige within the scientific community.[2] The Moore School had the feeling of a small, independent school within a university. There were about a hundred students, including graduate students, and fewer than a dozen professors.[3]

In June 1942, seven months after the United States entered the war and a few weeks before Kay and Fran were hired, Moore School dean Harold Pender and Colonel Leslie Simon, director of the BRL, held a meeting to discuss wartime cooperation on various projects, including establishing a satellite "computing group" at the Moore School.[4] A "computing program" in those days meant people calculating equations, often using electromechanical desktop calculators.

With so few men with mathematics backgrounds available to work as civilians, the two men discussed the Moore School assisting BRL by hiring and hosting women. Leslie wanted an initial group of about thirty-five. BRL already had an established computing group, and he would send some of his staff to train the new recruits.

Leslie and Harold agreed that Moore School's Philadelphia location was perfect, as it was located amid one of the densest collection of colleges in the country. The doors of the Moore School would open to a group of professional young women for the first time.

In July 1942, Kay and Fran entered the Moore School ready to start work. The young men they passed in the hallways looked at them and wondered what these young professional women were doing there. They were an unusual sight.

Inside their second-floor classroom, Fran and Kay found eight women and a few men, also part of the computing project, working at tables on big, clunky, metal, mechanical desktop calculators, the tool that aided their calculations.[5] Mechanical desktop calculators were more advanced than adding machines but were complicated to operate. They had gears that interlaced to perform an addition or multiple additions (multiplication). To use them, a person would enter numbers by pushing elevated buttons on the front, making a small clicking noise. In the 1930s, a special hand crank turned the gears for the chosen mathematical operation; by the early 1940s, electricity did it. The gears turning generated a loud and grating noise. Desktop calculators were heavy and expensive, costing thousands of dollars each in today's dollars.

"We were told we would have to do this computation of a trajectory," Kay remembered. Someone asked if she and Fran knew how to calculate a ballistics trajectory, and they said no, as it had not been part of any of their coursework at Chestnut Hill.[6]

In response, a man plopped down a "nice big fat book by an author by the name of Scarborough," Kay recalled, and said, "'Here is a chapter on numerical integration. Read that chapter and you'll know what to do.'"

Kay later shared, in her typically self-deprecating way, that she and Fran "read that chapter and were just as dumb as ever."[7]

Far from it being a matter of their intelligence, it turned out the man had given them *Numerical Mathematical Analysis*, a thick, dry academic treatise of Dr. J. B. Scarborough, which was intended for other PhD mathematical scholars, not math majors trying to learn how to use an advanced method of calculation called *numerical analysis*. He also knew quite well "numerical analysis" was a special technique then taught in graduate school but not yet part of any undergraduate's curriculum.

Fortunately, Lila Todd, a supervisor of the team, took the two women under her wing. "She was very patient and she taught us how to calculate a trajectory," Kay said.[8]

She and Fran were even unfamiliar with desk calculators, Kay recalled, "so now we learned desk calculators, and numerical integration, and trajectories, all in one fell swoop."

Lila Todd was a dynamo of a woman. Five feet tall and a 1941 graduate of Temple University, she had been the only female mathematician in a graduating class of 1,600 students. Lila was rarely encouraged in her chosen field—even the head of the Temple math department told her to choose another major—but she stuck with it.[9]

Following graduation, she joined the engineering department of DuPont, one of the world's largest chemical companies, based in Wilmington, Delaware. In March 1942, she began working for BRL. At BRL, Lila worked in the firing tables section until Captain Paul Gillon asked her to go to the Moore School to help supervise the many new women Computers they expected to hire.

During this time, even though she was a supervisor in Philadelphia, Lila had a "subprofessional" or SP rating as a civilian employee with the Army. She later said that "the [BRL] administration didn't believe that women should have professional ratings," no matter how well educated or how many people they supervised.[10]

Lila relocated to the Moore School along with another colleague, Willa Wyatt, where they were each responsible for a team.[11] Every day, they distributed individual sheets for ballistics trajectories to the Computers on their teams: formatted with columns and numbers representing the variables for each equation, including distance, humidity, winds, crosswinds, air density, the weights of the shells, temperature, and more.[12]

Kay, Fran, and the other Computers sat at individual desks with the desktop calculator on one side and the trajectory sheet on the other side, pencils in hand, ready to undertake the long process of calculation.

Each large white sheet of paper had a title at the top for the trajectory being calculated, including the type of artillery and projectile. Kay and Fran learned to work down the columns, creating row after row of calculations.

They used their desktop calculators to type in each number, and then add, subtract, multiply, or divide, as their calculations required. Their numbers represented the type of artillery being used, the speed and weight of its missile, and the temperature of the air, among many other factors. In doing these calculations, Kay and Fran were essentially pushing the missile's motion forward across its arc in the sky, step-by-step, to its explosive end at the completion of its journey.

Desktop calculators were cumbersome to use. For example, if Kay wanted to multiply 10,000 by 6, she had to hold "this little lever down six times" until she got an answer. It was a long process, but she admitted, "It was better than just multiplying by hand."[13]

There were plentiful opportunities for human error: mispunched numbers, miscopied results, and incorrect calculations made by using the wrong keys. Kay and Fran learned ways to review their work, and if they found errors, to go back and recalculate. Sometimes the machine itself would malfunction. The calculator would simply shut off midcalculation and they would have to reenter all the data.[14]

It took thirty to forty hours to calculate just one trajectory by hand. At the end of their shifts, Kay, Fran, and the other team members would hand in their sheets to Lila and she would lock the confidential military documents in a safe overnight. She returned them at the start of the next shift. This pattern would repeat itself every day until they could complete their work.

When a Computer finished, she handed the completed sheet back to her supervisor and got a new one from the seemingly never-ending stack. It was a lot of work for a missile that would reach its target in about forty seconds.

From the beginning, Kay and Fran were aware that, as women, they were oddities at the Moore School. Kay said, "[It] was just such a strange thing to the young fellows." Even though the women were a few years older, having already graduated from college, the young men "wanted to see what [we] were . . . We were strangers there." They would go out of their way to haze the women with juvenile pranks. For example, if Kay stopped to get a drink at the water fountain on the first floor, she could usually count on an engineering student to come along, twist the handle, and raise the level to douse her with water. He would crack wise that she had just received an "Ole Faceful," a play on the famous geyser at Yellowstone National Park. The men relished watching the women get a "face wash free from them," as Kay put it, a joke that wasn't very funny.[15]

Nestled in a Corner of the Base

The strings of the Army's Philadelphia Computing Section were pulled from the corner of a big Army base called Aberdeen Proving Ground.

When Kay and Fran started their jobs, the Moore School had hosted the Army's Philadelphia Computing Section for only a few months. The project recently had been relocated from the BRL at APG. *Proving ground* is the term the military uses to describe a parcel of land developed for the purposes of experimenting with or testing technology such as weapons or military tactics.[1] Congress had established APG in August 1917, just after the United States entered World War I.[2] It was located on 69,000 acres that jutted out into the Chesapeake Bay between Philadelphia and Baltimore.

Beginning in January 1918, APG tested weapons, including field artillery. Nestled on a small corner of the base was a group of mathematicians brought in to conduct research in the field of ballistics, or ways to get big guns to hit distant targets.

In the past, great gunners were expected to test the wind, judge the distance, choose the powder, and shoot their cannons. If they were skilled and lucky, they would hit the enemy's ship or troops before the enemy hit them. Job security and survival depended on master gunners' skills. But during World War I, new innovations in weapons technology necessitated a more team-based approach to firing artillery.

The new guns would shoot farther than a gunner could see on his own. Distances of five to ten miles were most common (e.g., 75-mm howitzers).[3] So teams were established that included forward observers who would set out on foot and by plane to see where the enemy lines were and locate targets for the artillery battalions. It was a dangerous job. President Teddy Roosevelt's son Quentin flew these missions as a pilot during World War I. He was one of the many young men lost in battle during the Great War.

The brave forward observers and pilots would communicate the target information back to their battalions by field telephone, by a fellow soldier, or by messages tied to dogs and carrier pigeons (carrier pigeons had a 95 percent success rate).[4]

But distance wasn't the only information crucial to the gunner's success. With targets at distances of many miles, weather in the field also impacted the trajectory. For example, if you shot an 18-pound shell a distance of five miles, then wind, rain, and even temperature could affect the arc. Like a good football quarterback judging the crosswinds in the stadium and distance to his wide receiver before throwing his long pass, the gunner needed a way to factor in conditions of the battlefield as he aimed and shot his artillery.

The Army recruited the mathematician Oswald Veblen (nephew of the great sociologist Thorstein Veblen) to help them develop an artillery trajectory strategy. Veblen earned his PhD in mathematics at the University of Chicago and since 1905 was a math professor at Princeton.[5] The Army created a small research division for him.[6] Their main mission was to answer the question, *Is there a way that mathematics can help the Army better aim its artillery?*

Veblen shared the overall goal of Aberdeen Proving Ground to come up with ways the Army could improve American artillery. At the time, Europeans were the leaders in the weapons industry and US troops largely used French and British guns.[7] The United States did not yet have the industrial capability and quality to build such high-precision pieces, but the folks at APG were dedicated to changing this. They became focused on the development and testing of new US artillery and other weapons and tools to support them, should there be another war.

Veblen had a clear objective for his program: calculating the trajectory of a missile from the time it left the muzzle of the cannon to the time it hit its target. This was not an easy problem, and Veblen tapped the services of friends and colleagues—like Dr. Gilbert A. Bliss of the University of Chicago—to help him.[8] Dr. Bliss, a professor of mathematics, helped pioneer a special area that he called the "mathematical theory of ballistics," or the effects of small changes on the path of a projectile (such as a missile) from wind currents, air temperature, speed, and weight.[9]

The set of differential calculus equations that emerged from Veblen's group's research would revolutionize the accuracy, and deadliness, of artillery. They could now determine with great certainty the angle the gunner should set the gun to hit a distant target. It felt like the golden answer they had been looking for, but there was one small problem.

Gunners could not do these complicated calculations in the field on their own—they took many hours and generally needed to be done by people with advanced mathematics skills. Plus, it wasn't a matter of just one calculation, but thousands, for the range of weather conditions the gunners might encounter on the battlefield. Wind, rain, snow, heavy crosswinds, light crosswinds, and temperatures hot and cold, all needed to be precalculated and compiled into "firing tables," reference tables that the artillery team could rapidly check the correct angle for aiming their cannons in the midst of a battle.

No existing machine of the day was capable of these calculations; only someone with graduate-level skills in certain types of differential calculus equations could do them. Veblen brought in a group of well-qualified young men—with mathematics degrees from institutions like the University of Chicago and Princeton—and began to hand-calculate ballistics trajectories.

When World War I ended in November 1918, Veblen returned to Princeton, yet still dedicated himself to continuing the task he had started. Although the United States and other countries disarmed and budgets were slashed after the "war to end all wars," he fought for continued funding to continue the Army's research on firing tables.[10] He traveled to Europe, even Germany, to learn how Europeans conducted their ballistics research and calculated their ballistics trajectories, and continued to advise APG on what he was finding.

In the meantime, the Army was developing new artillery, eventually to include the 105-mm howitzer, called "the workhorse of the artillery," and the larger, heavier 155-mm howitzer, nicknamed the "Long Tom."[11] (The Long Tom borrowed heavily from the French howitzer, the Grande Puissance.) Each weapon would need its own firing tables, which left Veblen shaking his head about the small staffing he had been given for his calculating teams.

In 1935, APG created the Research Division for this mathematical work, and in 1938 gave it a new name: the Ballistic Research Laboratory.[12] When the United States entered World War II in 1941, BRL was ready to ramp up its ballistics projects and firing tables. Veblen, now at Princeton's Institute for Advanced Study, was named the chief scientist of BRL and brought with him a newer and even more diverse group of mathematicians, engineers, astronomers, physicists, and other scientists to staff and advise BRL's staff on scientific and technical programs and problems. Among those involved were E. P. Hubble of the Mount Wilson Observatory, for whom the Hubble Space Telescope would be named; Subrahmanyan Chandrasekhar of the Yerkes Observatory; Isador Rabi, a Nobel Prize laureate in physics; Hugh L. Dryden, an aerodynamicist and key player in the founding of NASA; and John von Neumann, a world-famous physicist, who would play a key role in the development of the atomic bomb.[13]

Colonel Leslie Simon became director of BRL in 1941 and chose as his assistant director Captain Paul Gillon, a fellow graduate of West Point who had received his master's degree from MIT.[14] These two soldier-scientists shared a dedication to BRL's mission of applying science and mathematics to improving American ballistics, and they pushed a plan to make the artillery more accurate and the firing tables more accessible.

A key question facing Leslie and Paul was where to find the manpower to calculate all of the new trajectories. With so many men in uniform and overseas, the question soon changed to where to find enough "woman power" to calculate so many new ballistics trajectories.

If their search was limited to the immediate area around Aberdeen in rural Maryland, they would not find enough people, much less college-educated women with math degrees to run their complex differential calculus equations. Philadelphia, an hour and a half north, boasted Penn, Drexel University, and Temple University, all with female students. Then there were all-women's colleges in the area, like Bryn Mawr and Chestnut Hill. By June 1942, Leslie had met with Harold Pender at the Moore School to discuss establishing an Army computing group there.

But who would supervise them? Professor Bliss, one of the original members of Veblen's team, knew the perfect person for the job. Dr. Herman Goldstine was a former favorite student of his, an expert on numerical analysis who had just become a professor at the University of Michigan. Dr. Goldstine had also just been drafted, so Bliss wrote to Veblen and suggested that his former student go to BRL to continue his mentor's work.[15] After all, mathematicians were not needed on the front lines. Veblen quickly agreed and Herman was transferred to BRL.

When he heard about the reassignment, Herman was excited. After years of hearing stories about the Proving Ground from Bliss, he likely expected to spend the war deep in discussion with world-class scientists and mathematicians and long lunches at the base's Officers' Club with its big picture windows overlooking the Chesapeake Bay.

But Leslie and Paul had another plan in store for Herman. The newly appointed first lieutenant Herman Goldstine would soon head to Philadelphia to oversee the recruitment, training, and work of a growing group of young women to calculate ballistics trajectories in Philadelphia. Herman was not going to be happy when he found out.

Give Other People as Much Credit as You Give Yourself

A few weeks after Kay spotted the notice in the *Evening Bulletin*, a recent University of Pennsylvania graduate by the name of Betty Snyder saw a similar advertisement. When the United States entered the war, Betty had been working at a Philadelphia-based magazine, but now she contemplated how to take a more active role in serving her country because, as she put it, "everyone wanted to assist in the war."[1]

Even before the United States entered World War II, First Lady Eleanor Roosevelt had been seeking a greater role for women in the military. With Congress, she created the women's auxiliary army, which would grow to include the Women's Army Auxiliary Corps (WAAC) and the Navy's Women Accepted for Volunteer Emergency Service (WAVES).[2]

Betty had two sisters in WAVES, but when she tried to join herself, she was not accepted because she had crossed eyes. So she was delighted when she saw a notice in the newspaper about a job opportunity to help in the war effort at the Moore School.[3] Betty was quickly hired and, like Kay and Fran, told to report to a classroom on the second floor.

Frances Elizabeth "Betty" Snyder came into the world on March 7, 1917, the third of John and Frances Snyder's eight children, born in Philadelphia's Hahnemann University Hospital. She was raised in Narberth, the heart of Philadelphia's Main Line, which had a close-knit, small-town feel and a town center along the same cherry blossom–lined street as the train station. Every April, Philadelphians would flock to Narberth to see the cherry blossoms and crab apples in full bloom.

John was a teacher at Central High School for Boys in downtown Philadelphia, following in the footsteps of his father, Monroe Snyder. John was, by all accounts, a dramatic and engaging teacher. He would bring to life poems of Henry Wadsworth Longfellow for his children by standing near the living room fireplace and reciting "Evangeline: A Tale of Acadie" in his booming voice: "Still stands the forest primeval; but far away from its shadow / Side by side, in their nameless graves, the lovers are sleeping."

Betty's mother and father frequently attended the opera with the free tickets provided to teachers when the Metropolitan Opera came to Philadelphia. They always dressed in their finest evening clothes, made by Frances, because they didn't know where their seats would be, in the orchestra section right in front of the stage or up in the highest balcony, and they didn't want to be embarrassed if they were front and center. Sometimes John volunteered as a "super," or extra, acting on the stage in nonspeaking roles. In *Aida* he carried a spear. Inspired by her parents, young Betty listened to opera records and learned the scores. She particularly adored the famous Italian tenor, Enrico Caruso.

Monroe Snyder looked like Santa Claus but with a trimmer beard. Having graduated from the University of Michigan in 1872, he became an astronomy and math teacher at Central. Central was the first high school in the state to have its own observatory. Students would come at night to track the stars. Monroe became head of the "department of mathematics and astronomy" and director of the Observatory.

Before Betty's birth, Monroe was secretary to the United States Electrical Commission, and together with a group of scientists and technologists, he came up with the idea that a national bureau of physical standards was needed to ensure the products of electrical instrument makers and manufacturers were precise. The National Bureau of Standards was founded in 1901, and it was the first physical science research lab of the federal government. Monroe was also friends with Thomas Alva Edison and Alexander Graham Bell, inventors of the light bulb and telephone, respectively.

The Snyder home in Narberth was filled with books, inherited from one of Betty's uncles who had been an English teacher, so their library included classics by such authors as Charles Dickens and William Makepeace Thackeray, as well as the *Encyclopædia Britannica* and the *World Book Encyclopedia*. When she had to write reports for school, Betty had all the research she needed right at her fingertips. The family "was learning something all the time," Betty said. "We never had a day off if we weren't doing something. We didn't object... We didn't feel like we were being bossed around. It was just fun to learn something new."

Her parents were passionate believers in education for all of their children, boys *and* girls, and her father often took the children to the Free Library of Philadelphia on Logan Square. One day, Betty and her siblings researched sunspots in French and German. None of them knew the languages, but their father gave them words to look up so they could learn.

Retired by the time Betty knew him, her grandfather Monroe lived with the family from 1926 until his death in 1932, when Betty was fifteen. He shared her parents' passion for education and was always exposing them to something new, like the Greek alphabet. He dressed to the nines, often in a cutaway coat, starched collar and shirt, and striped trousers, carrying his books in a cloth bag. In the afternoons, he and Betty liked to walk into town, discussing chemistry, physics, and astronomy along the way. Though she often had no idea what he was talking about, she listened intently. When they reached the main street, Haverford Avenue, with its small shops, he would give her eleven cents. She would go off and return with a pickle in one hand (six cents) and an ice cream cone in the other (five cents). They would start the walk back home, with Grandpa Monroe continuing to wax philosophical as Betty took bites from the treats in both hands.

When she was ten and her brother Charles was seven, Monroe was having trouble with one of his desktop calculators, an Olivetti. Something inside it caused the divide circuit to malfunction. Betty and Charles took out every part of the machine, labeled it, and tried to figure out how it worked, but they couldn't fix the mistake. They put it back together again, and though it still didn't divide, they enjoyed their glimpse into it and were proud of getting all of the pieces back together. It was good to know how things worked.

Betty was not only cross-eyed but also left-handed and flat-footed. She had an infectious grin, long teeth, and glasses. She had to wear expensive corrective shoes. Her parents felt they had to look after her more than they did their other children. "I think that they treated me a little bit sweeter," Betty said. "I think maybe that made a difference. I think [the vision and foot problems] also helped me too. I think it made me work harder."

On Sundays, her father, John, a gifted speaker, gave talks at different churches. One Quaker meeting liked him so much that they offered him a scholarship so that one of his children could attend a

Quaker school. A boarding and day school in Pennsylvania, George School was about forty miles from Narberth. Fair-minded, John offered it first to his oldest child, Betty's sister Eleanor, and then to Betty's older brother John. But Eleanor didn't want to be apart from her friends and John didn't want to go either. As soon as John extended the offer, Betty leaped at the chance. She would go!

Before she left, her grandfather told her, "If you really want to get ahead in the world, be honest and don't take credit for what other people have done. Give other people as much credit as you give yourself." She would remember that forever.

Betty loved George School. Her classes were long, an hour and a half each, with only seven students per class. In Quaker tradition, they were coed. She learned Latin, French, English, the classics, algebra, geometry, solid geometry, and trigonometry. She loved mathematics and in her senior year took a side job correcting geometry papers for a teacher for the grand sum of $100, which she considered a fortune. She played sports, including field hockey, and also the bass viol (a six-stringed instrument of the Renaissance and baroque periods), trumpet, and piano.

When it came time to graduate, Betty received two college scholarship offers: the University of Pennsylvania and Temple University. She chose Penn because it was practical: For one, her brother John "was taking his PhD there and he drove a car. It was during the Depression, and I could get to college by his car."[4]

Professionally, she was clearer on what she did not want to do than what she did. Her older sisters attended secretarial schools, studying shorthand and typing. Betty knew she did not want to be "a secretary, or keypunch operator." Keypunch operators, mostly women, spent their days punching holes in cards to store data; the cards were fed into machines that could read the data and use it for calculations and creating reports.

At the time she entered Penn in 1935, it had just become coeducational. The College of Liberal Arts for Women had been founded only two years earlier, offering women a full-time, four-year, liberal arts undergraduate degree for the first time in Penn's history. But the faculty was still largely male, and most classes were separated by gender. This was a surprise to Betty, who expected coed classes in a university that claimed it was coed, as George School did.[5]

By 1940, a year after Betty graduated, women made up about a quarter of Penn undergraduates, about 200 out of 800. An edition of *Women's Undergraduate Record* from that period had this account of being a freshman woman at Penn:

> [The male upperclassmen] called us "green" and gave us green name-buttons to wear. We laughed with them about it, but secretly we felt that we were at least a little wiser than the upper-classmen thought we were. We didn't press the point, however. We were too busy learning to walk in this new college world... We were the first class to be orientated instead of hazed, an innovation for which we were especially grateful.[6]

Based on a test that Betty took in high school, the George School advisers recommended she major in mathematics in college. "I agreed with them," Betty remembered. She signed up for calculus in her first semester. On the first day of her first math class, she had been playing hockey by the Schuylkill River and ran all the way to class to make it on time. As she took her seat, she saw a male, elderly mathematics professor at the front of the room who glared at her and the students in the room, all women, and roared: "You women should be at home raising children!"[7]

He said that to them at the start of every class for the entire semester. At George School, teachers treated students with respect and an egalitarian attitude. Betty was dismayed, but decided to wait it out until the next semester.

When it was finally time to pick the next semester's classes, Betty was determined to enroll in a new math class with a new teacher. But when she checked the roster, she found only the same professor listed. He was the only teacher that taught women's math classes. Betty decided she could not stand that abuse again and switched majors. She chose journalism, which allowed her to take classes across several departments of the university. She took as many credits as her advisers would allow and finished about five years' worth of classes in four.

At Penn she was active in the German Club and the dramatic club called the Penn Players. She had a close-knit group of female friends—they were all serious, smart, hardworking. Many of her friends were working for their PhDs, but Betty felt that a college degree would be sufficient to get a good job and earn a salary.

She graduated in 1939, in the tenth year of the Great Depression. She wanted to work in journalism and secured a position at *Farm Journal* magazine. Philadelphia-based, it had first been published in 1877 as a resource for farmers in the agricultural areas close to the city. Its founder, Wilmer Atkinson, like Betty, was a Quaker and believed in distributing common-sense information to farmers and their wives. By 1915, *Farm Journal* had a million subscribers nationwide.[8]

Betty's job was in the statistical department. "I thought if I could get in the back door, then maybe I could write for the *Farm Journal*," she said. She was made supervisor when her manager left on maternity leave. Soon after, *Farm Journal* decided to conduct a survey for advertising purposes, to find out how many lipsticks farmers' wives bought. *Farm Journal* staff conducted the survey and the statistical department ran the results. They came back showing that the poorest farmers' wives, the ones with no indoor plumbing, bought ten lipsticks a year.

Betty was surprised. She was a young professional woman, wearing lipstick to the office every day, and she bought only one or two lipsticks a year. Surely farm women during the Depression, who had to stretch their scarce money across families and farms, would not buy more. Logically, it did not make sense.

Not wanting to contribute incorrect data to a story, Betty summoned her courage and marched into her boss's office. She asked him to rerun the survey analysis. It would be a time-consuming proposition, but she thought it was important for the integrity of the magazine's reporting. Her boss, a PhD economist on leave from Penn, thought about her idea and eventually agreed. He hired extra keypunch operators to rerun the data quickly, as publication deadlines were approaching.

Betty turned out to be right. The problem tracked back to a mistyped punch card. On the first run, a keypunch operator had "offpunched" the data, meaning that she punched a 1 into the tens column rather than the ones column of the punched card in the row representing lipsticks purchased.[9] (The thin cardboard punch cards had twelve rows and eighty columns each.) This was a common error but one with serious consequences, especially had Betty not seen the problem. The *Journal* ran its story with the correct results, and Betty learned to trust her logic. It would become one of her most valuable life skills.

Though she enjoyed her job at the magazine, with the onset of the war, Betty was looking for an opportunity to serve her nation. The front-page Army job notice beckoned to her, and Betty joined the Army's Philadelphia Computing Section on August 19, 1942, at the age of twenty-five.[10]

By the time Betty joined, a month or two after Fran and Kay, the dean, John Grist Brainerd, had enlisted an octogenarian Penn math professor to teach numerical analysis. It was a repeat of Betty's earlier experience. Although this professor did not yell at his students, he clearly did not like teaching women, turning his back to Betty and the other Computers and lecturing to the chalkboard.[11]

Betty and the other students knew they were not learning what they needed to know. Eventually they told their supervisors that to perform trajectory calculations for the Army, they needed a teacher who would teach to and engage with them. It was a huge problem.

Grist sent flyers to the American Mathematical Society (AMS) and the American Association of University Women, and sent letters to the Penn faculty asking them to "volunteer their daughters or their daughters' friends."[12] They hoped their outreach would work.

We Found Things in a Not Very Good State

On September 1, 1942, Captain Paul Gillon and Herman Goldstine traveled to Philadelphia to review the women's computing project. As Herman put it, "We found things in a not very good state." This project, like many new ones, suffered from growing pains and needed more people, more desktop calculators, and more support. Also, it was clear that the elderly Penn mathematics faculty did not want to work with young women. Overall, the women's computing project "was in need of leadership."[1]

At the end of September 1942, Leslie and Paul placed Herman in charge of BRL's women's computing project in Philadelphia, and he and his wife of one year, Adele Katz, moved to the 260-year-old city. It would be up to Herman to figure out new ways to speed up the calculation of trajectories so that BRL could produce new firing tables. This job was going to be a headache!

It was already clear that Herman needed better ways to recruit young women and to train them quickly for the differential calculus calculations they would need. It turned out he knew just the person to lead: his own wife.

Adele was twenty-two, a Brooklyn native, and a graduate of Hunter College, New York's all-women university. In a remarkable decision for her time, at only twenty years old she decided to move 600 miles, on her own, to study graduate-level mathematics at the University of Michigan. Sophisticated and exuberant, Adele was not only a great mathematical mind but also a skilled violinist.

She met Herman as a graduate student, received her master's in mathematics, and was on her way to a doctorate before Herman was drafted and their lives (along with millions of others) turned upside down.[2]

Herman got Grist to terminate the Moore School's arrangement with retired male math professors because he knew who should teach instead. At his request, Moore School appointed Adele and together they hired several women to teach the courses for incoming Computers.[3]

Knowing the Army needed to hire broadly to recruit enough Computers, Adele created two instruction tracks. A shorter course for new Computers who already had mathematics degrees and a longer one for those who needed a background in calculus before learning graduate-level numerical analysis techniques. The latter course would open the doors to a wide array of new recruits for the Army's Philadelphia Computing Section.

Adele taught her first students in the fall of 1942. Her teaching style was unforgettable to many of her students, who, years later, remarked on both her grace and her brilliance.[4] She would walk into the classroom smoking a cigarette and would elegantly sit down on the edge of the desk facing them, her legs neatly crossed. She had upswept hair and wore a tailored shirt and skirt. With this well-dressed woman lecturing easily on differential calculus and numerical analysis, her students felt inspired and excited.

Adele taught her students to calculate the flight of a missile down to every tenth of a second. With enough accurate samples, they could map out the arc of the missile's path from the gun to its target by calculating specific points along its path.

Adele was happy. The teaching job, she felt, fit her "like a glove, since it involved teaching serious, adult students who wanted to learn."[5]

Adding Machines and Radar

In the fall of 1942, soon after the Goldstines moved from Michigan to Philadelphia, a twenty-year-old recent Temple University graduate named Marlyn Wescoff had been feeling frustrated in her job search. A recent graduate of Temple's Secondary Education Department, she wanted a good job with a good salary. She had a sweet, upturned mouth, well-defined cheekbones, and luminous eyes that bore a perpetually bemused expression. Although shy, she reached out to her network of family and friends. One day, she received a call from a friend of her sister's asking if she knew how to run an adding machine. She did.

"Well, at the University of Pennsylvania they're looking for people who can run adding machines. Why don't you go up there?"[1]

Marlyn went to the Moore School and met Dr. John Mauchly and his wife, Mary Walzl Mauchly. The Mauchlys were working on an Army project at the Moore School conducting radar experiments. John performed radar experiments on the roof with students, and Mary, trained in mathematics, supervised the calculations based on these experiments on the second floor.

In the fall of 1942, Marlyn showed up to interview with Mary in a second-floor classroom of the Moore School Building.

"Do you know how to use a calculating machine?" Mary asked her.

"I don't know what that is," Marlyn answered. "I can use an adding machine."

"If you can use an adding machine," Mary told her, "I can show you how to use a calculator." She explained that the desktop calculator had a few more functions than the adding machine, including multiplication and division.

For fifty cents an hour (about $4 a day, or $1,000 a year), Marlyn became the only full-time employee doing calculations for Mary and John. She worked alongside male students of the Moore School who popped in and out for shorter calculating stints. Mary would give Marlyn sheets of papers with equations worked out based on the radar data provided by John's experiments. Marlyn, for her part, worked her way down the sheets, checking calculation by calculation, then handing each one back to Mary when she was done.[2]

Marlyn Wescoff was born on March 2, 1922, the younger of two daughters of Jewish parents who lived in West Philadelphia. Her father, Fred, was a traveling salesman who carried sample cases for a dress manufacturer. Fred was born in New York, and her mother, Anne, was born in Russia and immigrated at age three. Anne never talked about Russia.

Marlyn never knew that her family, like so many others during the Depression, had money problems. She remembered they "always lived in a nice house, and we had plenty to eat. We didn't have anything else." For school, she and her sister each had two sets of clothes, and they would wear one set while their mother washed the other. It did not seem out of the ordinary; few in her diverse Irish, Italian, Jewish immigrant neighborhood had more.[3]

Like Betty's, Fran's, and Kay's parents, Marlyn's parents were deeply encouraging of Marlyn and her sister Charlotte's educations. Their house was full of culture—music and books, especially. Marlyn's mother bought a secondhand piano, and Marlyn learned to play from her aunt, the neighborhood piano teacher, and her sister.[4]

But shy Marlyn loved books most of all. Her parents bought used collections, and she feasted on Charles Dickens, Edgar Allan Poe, and Guy de Maupassant, the nineteenth-century French author. To burrow down in a cozy nook and escape into a book was her favorite activity in the evening after homework and on the weekends.

Every Saturday in high school, Marlyn and her friend braved the crowded, noisy streets of Philadelphia, riding the trolley to the great downtown main library to collect more books. The huge Philadelphia main library anchored a free library system founded by Benjamin Franklin in 1731 that spread across Pennsylvania and the country. With hundreds of thousands of volumes to borrow, there were more than enough classics to choose from, and each week Marlyn left clutching her next six books, the maximum. She returned the next week, every book read, ready for more adventures.[5]

Student tickets to the Philadelphia Orchestra at the Academy of Music rounded out Marlyn's music education, and she went often with her sister. The orchestra, conducted by Leopold Stokowski, would play the soundtrack for Walt Disney's *Fantasia* when it came out in 1940.[6]

Marlyn's family was interested in politics, history, unions, and news about the coming war in Europe. At the dinner table, they talked politics every night. Everyone read the daily newspaper and discussed world events. Then the family adjourned to the living room and tuned their large wooden radio to shows that brought the music of the Philadelphia Orchestra or the comedies of Edgar Bergen and his famous dummy Charlie McCarthy into their homes.

At West Philadelphia High School, Philadelphia's first secondary school west of the Schuylkill River, Marlyn studied history, French, and English. "Science and math frightened me, although I did take math courses," Marlyn said. Like Fran, she graduated high school at sixteen, after skipping several grades.

She entered Temple at "a very immature sixteen" and feeling at "a disadvantage socially" among the worldly eighteen-year-olds. But she found some friends of the same age and they "stuck together."

At Temple, Marlyn attended classes in secondary education and majored in history and sociology. She added a minor in business because "that was my ace in the hole: If I didn't get a job as a historian, I could type and do shorthand."[7]

Marlyn had been in the middle of her junior year when the United States entered the war. "Everybody was doing what they could," she recalled, "and jobs were opening up, people were working more, and the city was growing, and there was more hustle and bustle."[8]

She became a furious knitter—making scarves, helmets, and gloves for the American Red Cross. Helmets made for cadets were not only tangible forms of war support but they also enabled women to feel they had an active part in the conflict. Even before Pearl Harbor, Americans were knitting and sending care packages called "Bundles for Britain." Many women proudly carried around their knitting needles—even First Lady Eleanor Roosevelt was often photographed knitting or carrying her knitting bag. Marlyn, however, did not have much confidence in her knitting abilities and imagined that "somebody back at Red Cross headquarters" was ripping her stitches out and redoing her work.

The summer before her senior year, Marlyn did her student-teaching hours in a Philadelphia high school, even though she was barely older than her students. She was still shy, and it was hard for her to be in front of a classroom, but it was a good career path, so she did it and then finished her requirements for graduation and teaching certification. As graduation approached, Marlyn prepared to apply for teaching jobs.

The night before graduation, there was a supper for graduating seniors and their families. The dean stood up and made a startling declaration. "Those of you who are Jewish," he said, "don't look for any jobs in the suburbs. Don't look anywhere outside of Pittsburgh or Philadelphia. Nobody will hire you." To teach in Philadelphia would also be difficult, he explained, because the older teachers were entrenched and weren't leaving. He said all this matter-of-factly. This was the truth, and he only wanted them to be realistic.

For Marlyn "that was sort of a shocker, at that date in my career, right before we graduated, to be told that." But it was not the first time her family had encountered anti-Semitism. When she was in elementary school, her parents went off on a trip to Lake Hopatcong in New Jersey, a freshwater lake and tourist attraction near the Delaware Water Gap. Marlyn was surprised when her parents arrived home early, within a night or two. Confused, she asked her mother what had happened. Anne began to weep. It turned out that every hotel and lodging house had a sign reading NO JEWS OR DOGS ALLOWED.[9]

After the dean's words, Marlyn did not try to apply for jobs as a teacher. She spent the summer in a rather shocked limbo and then applied to be a secretary at a longshoreman's union in nearby Camden. She didn't get the job. But soon the position with Mary and John Mauchly would put Marlyn's life on a new course.

Working for the Mauchlys, Marlyn soon gained a reputation for never making a mistake in her calculations. She could run them all day, typing numbers into the desktop calculator with its clicking keys and excruciatingly loud gears, writing down the results. For many other people, errors were natural in this process, but not for Marlyn. She did not seem to notice, but others did.[10]

Marlyn did not know details about the radar experiment, beyond that antennae were being tested on the roof, and she did not ask the Mauchlys or anyone for more information. During the war, people were careful not to ask questions about confidential military projects, and signs everywhere suggested that sharing confidential data improperly could endanger soldiers and others. LOOSE LIPS MIGHT SINK SHIPS, warned one of the most famous posters of the time.[11]

In the hallways of the Moore School, Marlyn saw other women and men, but her interactions were limited to her own team.[12] That suited her fine, because she liked the Mauchlys. Marlyn's temperament resonated particularly strongly with John's. She was shy and he, too, was soft-spoken and mild-mannered with a rumpled appearance. Yet he was responsive to everyone, men and women alike.[13]

John William Mauchly, known as Bill to his parents, was born in Cincinnati on August 30, 1907, the elder of two children of Sebastian and Rachel Mauchly. When John was a young child, they moved to Chevy Chase, Maryland, then a suburban area that still had a rural feel. Sebastian raised poultry and liked living off the land. John had to clean the chicken coop, which contained fifty chickens, every Saturday. He hated it.

Sebastian was a physicist with the Carnegie Institution of Washington's Department of Terrestrial Magnetism and researched Earth's magnetic field and the way lightning works. Many scientists lived in Chevy Chase, including officials from the National Bureau of Standards and the National Weather Service.

John had loved science from an early age. On the Fourth of July, he would fix up a contraption and fireworks would go off fifty feet away. On April Fool's Day, he wired the front doorbell of the family's home so visitors got a shock. He read voraciously, consuming books and *Popular Mechanics*.

In high school, he was an ace at math and physics. He belonged to the National Honor Society, debated, and edited the school newspaper. He attended Johns Hopkins University in Baltimore on a state scholarship to study engineering but by sophomore year was uninspired and the school let him transfer his scholarship to physics.

In 1928, John graduated from Johns Hopkins, and in 1930, he married Mary Augusta Walzl. A mathematics graduate of Western Maryland College, Mary was the oldest of three children, with two younger brothers. Her mother was from a Roman Catholic family in Maryland and her father was a photographer. She announced at six that she would have nothing to do with Roman Catholicism. Her parents died when she was a child, and she was raised by relatives in Maryland. She began dating John while attending Western Maryland College and they married after she graduated.

At Johns Hopkins, John entered into the graduate physics program, studying molecular spectroscopy, or the investigation and measurement of spectra produced when matter interacts with or emits electromagnetic radiation. This meant he spent countless hours calculating the molecular energy of gases. He received his PhD in physics in 1932 and found the Depression-era job market to be challenging. His doctoral degree (he never bothered to collect his bachelor's or master's degrees) was in an unpopular field, and most institutions did not have money to spend on the research.[14]

He stayed at Johns Hopkins as a research assistant to one of his professors, doing calculations on a Marchant. The following year he received an offer from Ursinus College to head the physics department. Ursinus was then a sixty-three-year-old liberal arts school twenty-five miles from Philadelphia, on a campus shaded with trees, in a town called Collegeville, with a population under a thousand.

When he came to Ursinus, it had no physics department, no physics professor, and awarded no physics degrees; it only offered the subject to premed students who required it.[15] To attract students to his courses, John built up the department and tried to open the minds of those who did come to other career paths besides medical school and teaching.[16]

In John's Christmas lecture, which he gave on the last day of classes before winter break, he used the laws of physics to gain the clues to a Christmas package's contents: measuring, weighing, submerging in water, and poking it with a long needle.[17] It grew so popular that soon other professors would dismiss their students, who were usually hungover from partying the night before, so they could attend John's lecture. It was so well attended that it was usually held in the auditorium, and professors would come and listen too.[18]

For his lecture on Newton's laws of motion, John built a skateboard by affixing roller skates to a board. He skateboarded into the classroom, climbed up on the lab table, and demonstrated the forces of motion as he rode on the table.[19] If he moved to the right, the board moved to the left, and vice versa—an equal and opposite reaction. In another lecture, he arrived on roller skates and performed spins, moving his arms out to slow down and folding them in to start rapidly spinning, much to the delight of his students, who never forgot their lecture on momentum.[20]

Marlyn worked for Mary and John Mauchly for about seven months. Then, in the spring of 1943, John told her that his radar project was being disbanded. He suggested she try to get a civil service job as an Assistant Computer for the Army computing project. He promised to put in a good word for her, without adding that he already had.

3436 Walnut Street

In spring 1943, Marlyn joined Fran, Kay, and Betty as an Assistant Computer for the Army's Philadelphia Computing Section. Marlyn took Adele's course for nonmath majors. The class went on for about three months, starting with calculus, then numerical analysis, and finishing up with calculating techniques for ballistics trajectory equations. Additional teaching staff included Mary Mauchly, who Marlyn knew well, and Mildred Kramer.

The class was hard for Marlyn, who had never taken calculus, but she gave it her best shot. Her notes, taken on the back of an old American Association of University Women recruiting flyer and treasured for over fifty years, show the sophisticated concepts Adele presented and discussed:

1. Interpolate
2. Integrate
3. Smooth—graduation—smooth function using differences
4. Solution of systems of equations
5. Graphing—Dive Bombing
6. Conversion of tables
7. Divided differences of D.B. checking

8. Differentiation—on D.B.—½ sec diff, ⅓ 2nd—partial derivative diff with respect to 1 or 4 variables
9. Compute + use graphical tables.[1]

Marlyn's head often spun from the classes, and she was very tired at the end of the day, but she kept going and finished along with the other women. She worked hard, and in the end found the course wonderful, strange, and difficult. Adele and Mildred did their job well.[2]

After three months, Marlyn was told to report to the computing group on the third floor of 3436 Walnut Street, across the quad from the Moore School. There were two attached rowhouses and 3436 was one of them. A sign on the door read WOMEN ONLY. Marlyn was told never to tell anyone outside of the project that she was working for Aberdeen Proving Ground.

As planned, the Army had taken over several buildings on the Penn campus separate from the Moore School where Fran, Kay, and Betty had first worked. There simply was not enough room for the dozens of women that BRL was hiring. Six sections of women would work at one such commandeered residence, 3436 Walnut Street, one on each floor, three on day shift and three on night. It was a plan designed to produce trajectory equations at an unprecedented pace.

When Marlyn first entered 3436 Walnut Street, she saw a long, rectangular room. Women in two rows sat at eight desks, hard at work on eight desktop calculators. With a bit of trepidation, she walked through this room of strangers to the long, narrow staircase and proceeded up to the second floor, where another team was arranged in a similar way, and then finally up the last staircase to the third floor where two rows of women awaited her.[3]

On the third floor, her new supervisor, Florence Gealt, welcomed her warmly. The women in the room waved and smiled, then quickly resumed work. Introductions and casual chat would have to wait for a break. Marlyn did not know it then, but the women around her would become good friends and the "Third-Floor Computing Team" would form a tight bond.

Florence handed Marlyn the first equation—on a long, white, oversized piece of paper—and pointed to an empty desk and waiting desktop calculator. After Marlyn finished one sheet, she began another. Marlyn soon learned that some trajectories required more calculations than others; some took a few days, others close to a week.[4] On average it took about thirty hours to calculate a trajectory using a desktop calculator.

As with other computing teams, Marlyn's third-floor group worked either the day shift, from 8:00 a.m. to 4:30 p.m., or the night shift, from 4:30 p.m. to 1:00 a.m., with a half hour off for lunch or dinner and ten-minute breaks as needed.[5] They worked two weeks on the day shift and then two weeks on the night shift, a regular cycle.

The third-floor room had no air-conditioning and in the hot, muggy Philadelphia summer, that was hard. Marlyn remembered, "It was very hot in the summertime, so they brought in a fan to help us out, and a water cooler."[6] On the watercooler during the summer was a big jar of salt tablets, which their senior civilian supervisor, John Holberton, provided to help prevent dehydration. The bathroom held a large jar of aspirin tablets for when headaches from the clicking and clacking desktop calculators grew intolerable, and a cot to lie down for a quick rest.

A polite gentleman who had grown up on his family's farm in southern Virginia, John Holberton received his degree in physics from College of William & Mary, and his master's in physics from Temple University. Initially, John Holberton became a junior scientific aide at the National Advisory Committee for Aeronautics (a forerunner of NASA) at Langley Research Center in Virginia. Then in 1937, he took a position at BRL as Senior Computer in Ballistic Computations.[7]

When the Army relocated much of the project to Philadelphia, BRL transferred Holberton to the Moore School as a civilian supervisor to help Herman run the Army computing project. Soon he became a favorite person to the dozens of female Computers.

Holberton supervised all the computing teams and, by most accounts, was very good at his job. The women found him easygoing and pleasant. He had "a nice word to say to everybody whenever he came up," Marlyn recalled.[8]

If John Holberton's manner was kind and attentive, Herman's was cold and aloof. When he visited the Third-Floor Computing Team, he would rarely even say hello and did not give much encouragement to the women on his teams. He tended to scowl, and when he arrived, silence would fall across the room.[9]

Herman would have a few necessary words with Florence and then leave the third floor abruptly—no good-byes. In fact, the only sounds were sighs of relief from the Computers.[10]

Though there was little time for chitchat during the day, the Third-Floor Computing Team had much to say and share both before and after working hours. Marlyn became fast friends with the other Computers on her floor, such as Shirley and Doris Blumberg, identical twins who had gone to the highly prestigious public school Philadelphia High School for Girls, or Girls' High.[11] And a few months after Marlyn arrived, two women had come down from New York City to join the group: Ruth Lichterman and Gloria Gordon. Then there was Grace Potts, who was from Reading, Pennsylvania, and had a BA in psychology.[12] Finally, there was Alyce Hall, an African American woman who was a married mother of two and a former schoolteacher.

Born Alyce Louise McLaine in 1908 in West Philadelphia, Alyce was the second of seven daughters. Her father, Smith, was a property manager and her mother, Catherine, taught at Fanny Coppin's Institute for Colored Youth in Cheyney, Pennsylvania. The family moved to Bryn Mawr, Pennsylvania, when Alyce was a child. After Catherine married Smith, she was not permitted by the school to continue working and became a laundress and homemaker.[13]

Alyce graduated Lower Merion High School in 1927. Like five of her siblings, she attended college, graduating in 1929 from West Chester State Teachers College, located about twenty miles west of Philadelphia. The school admitted African American students but denied them on-campus housing and dining room privileges. African American students had to live with African American families in West Chester.

Alyce began teaching in the Darby Township, Pennsylvania, schools. She was known as a creative instructor, teaching her students to knit and crochet and then demonstrating the math behind their crafts.[14] In 1932, she married a man named Marvin Hall, and after the United States entered the war, she taught math at Aberdeen Proving Ground and later was sent to the Moore School.[15]

As the Computers toiled away, smoking and sweating, supervisors told them that their tables were being sent to North Africa. There was a sense of urgency about their calculations; Marlyn remembers that the Army was "trying to get us to speed it up."[16]

One reason is that the BRL had to recalculate a bunch of firing tables, and quickly. It turned out that the tables the Computers had been making for Europe did not work in Africa because the ground was softer, causing the guns to act differently.[17] The allowances that Army ordnance had made for the recoil effect, or the kickback when the cannons were fired, were inadequate. Artillery recoils lessened a shell's velocity and changed its tilt, which could cause a gunner to miss his target. In addition, the air over equatorial deserts had different temperature and density than the values normally used in firing tables and calculated based on normal conditions in the United States.[18]

The North African campaign had started in the summer of 1940, as Axis and Allied forces pushed each other in the desert. In September 1940, Italy invaded Egypt, and in a counterattack that December, British and Indian forces captured more than 100,000 Italians. In response, Adolf Hitler sent in the Afrika Korps, headed by General Erwin Rommel, the great "desert fox." In late 1942, the battles continued, back and forth across Libya and Egypt. The pushes reached a pitch during the Second Battle of El Alamein, when the British Eighth Army drove Axis forces from Egypt to Tunisia. In November 1942, thousands of American and British forces were sent across western North Africa to join the attack. General George Patton would be assigned to lead them in March 1943.

The Computers knew few details of the battles or of the firing tables' intended usage, but they did know that eventually their calculations were reaching the front lines of the battlefields and having a major impact. They sometimes wrote personal notes of support to the soldiers in the margins of their firing table sheets but never heard a word back. This is because their trajectory sheets went down to APG to be compiled into firing tables and were only sent in that summary form to the troops. They never received the Computers' notes, and the Computers were not aware the soldiers were only receiving a summary of their firing table work.

By 1943, the Army's Philadelphia Computing Section was a success and the women's calculations for the BRL firing tables had proven their value: The Army would not ship artillery to the battlefield without its firing table. But this created its own problem. As the Army kept improving its arsenal with new artillery and major improvements to older artillery, it needed new firing tables with hundreds of calculations each. BRL, even with the calculating teams in Aberdeen and Philadelphia, was not keeping up. Despite their successes, Herman was upset, as valued artillery sat unused. What more could be done to speed calculations?

In the spring of 1943, Adele went on a recruiting trip to try to hire more college graduates. In conjunction with the American Association of University Women, she visited Bryn Mawr and Swarthmore outside Philadelphia; Goucher College in Baltimore; Douglass College in New Jersey; and Hunter College and Queens College in New York City.[19]

In New York City, on Monday, June 14, 1943, an article appeared in the *Brooklyn Daily Eagle* titled, ARMY NEEDS MATH MAJORS: ORDNANCE DEPARTMENT SEEKS COLLEGE WOMEN; MOORE SCHOOL WILL TRAIN NEW RECRUITS.[20] "A number of college women, preferably those of a mathematical bent," it began, "are urgently needed by the Army Ordnance Department for extremely important war duties." The article said that a "representative of the Moore School" (Adele) was available to discuss the course with interested college women. The course would train women "to carry on the work of the Ordnance Unit" but would also offer "broad training in college mathematics."[21]

From New York, Adele wrote Grist that she did not expect much because "next week is exam week."[22]

But she was able to reach a bright young woman at her own alma mater, Hunter College, whose father, Simon Lichterman, a Russian-born, Bronx Hebrew school teacher, saw the article and told his only daughter, Ruth, about it.

Ruth Lichterman, nineteen, was a quiet, deliberate, beautiful woman who had just completed her sophomore year at Hunter, intending to major in math. She had taken her final exams for the year and was planning to work as a waitress at Camp Copake in Columbia County, New York, an adult summer camp for young Jews from the city.[23] But when Simon saw the article, he and Ruth felt the job fit her perfectly, and she began to think she might change her plans.

She applied and was accepted, but the notice had not specified Philadelphia.[24] When the Lichtermans found out, they were devastated, not wanting Ruth to live in another city, even though it was only a hundred miles from New York. Still, the pay was too good for Ruth to turn down—$40 a week was more than Simon's teaching salary. Everyone was trying to "do their part," and if this was the way she could do it, even if it meant leaving the world she knew, she would try.[25] Soon she had moved to Philadelphia, studied with Adele, and taken her seat on the third floor of 3436 Walnut Street, where Marlyn, the Blumberg twins, and the others greeted her with warmth.

Ruth Lichterman was born on February 1, 1924, to Simon Lichterman and Sarah Schreibman. Simon had emigrated from Vilna, Russia, in 1913 at the age of thirteen and was such a good student that he was accepted into the electrical engineering college Cooper Union in Greenwich Village (Cooper Union served as a magnet for bright immigrant children interested in studying engineering or art because tuition was fully paid from an endowment left by Peter Cooper, who designed and built the first US steam locomotive). However, it was the height of World War I, so instead of attending, he enlisted in the Army at age seventeen and was sent to Panama, Britain, and Palestine.[26]

Ruth's mother, Sarah, was also born in Russia and had emigrated to the United States using the passport of a male cousin and dressed as a boy. She took a job in a dress factory, and the family eventually saved enough money to pay back the cousin.[27]

When Simon returned from the Army, he planned to finish Cooper Union and become an electrical engineer. But his brother, Irving, who had also returned from the Army, had obtained a job teaching Hebrew school. Irving was making a good salary, and Simon decided to become a Hebrew school teacher himself. He met Sarah in a political club and fell in love with her instantly, but she had eyes for a different man closer to her age—Simon was seven years younger. The object of Sarah's affection turned out to be afraid of marriage, but Sarah held out hope. Simon courted Sarah for three years before she finally married him in December 1922.[28]

Ruth was born fourteen months later. They lived in an apartment in the Bronx, and after Sarah's sister, Freda, saved enough money to emigrate, she joined the Lichterman household. Ruth shared a bedroom with her; her parents had the other bedroom. The apartment building was across from a school and a park. Children played outdoor games like Ringolevio and stayed out late, which was typical for the time.[29] Ruth's younger brother, Herbert, was born when she was five but died of rheumatic fever five years later. Changed forever by her grief, Sarah became deeply critical of Ruth, bossy and so difficult that Ruth's friends did not like her.

Before Ruth entered high school, her grandmother died. They moved to a one-bedroom apartment, also in the Bronx. By this time, Freda had married and moved out, and Ruth slept in the living room alone, an improvement. The apartment was across the street from Ruth's high school, Morris High School, which was at Boston Road and East 166th Street in Melrose, the Bronx. It was the first public high school in the borough.

Ruth was bubbly and talkative, with expressive eyes and an elegant bearing. Boys were always drawn to her. Personable as she was, though, she also had a serious, practical side. She met a boy in Simon's Torah class, but he was a college student so she decided he was too old for her.

She entered Hunter College in the fall of 1941. The following summer she took a waitressing job at Camp Copake to have fun and save money for tuition. But she turned down the opportunity to go a second summer and moved to Philadelphia, putting college on hold and supporting her family and nation.

Assigned to the Third-Floor Computing Team at 3436 Walnut Street, Ruth worked four desks up from Marlyn. She enjoyed calculating trajectories, which she likened to doing a puzzle. Around this time, the whole team became close. Many days, before the night shift, they met and had dinner together. Ruth and the others often went out on the town to restaurants, bowling alleys, or movie houses. In 1943, the films *Casablanca, For Whom the Bell Tolls, The Song of Bernadette* (starring Jennifer Jones), *Shadow of a Doubt, Edge of Darkness, Mission to Moscow, So Proudly We Hail!, Stormy Weather, Phantom of the Opera*, and *Lassie Come Home* were released. Center City featured the best-known movie theaters in Philadelphia, on Market, Chestnut South, and North 8th Streets, with millions of spectators going to the movies. There was the Boyd Theatre, an art deco palace; the Stanton (with almost 1,500 seats); the Stanley (nearly 3,000 seats); the Aldine; and the Karlton (about 1,000 seats), which featured marble, murals, and gilding.

Mastbaum Theatre at 20th and Market Streets was one of the largest in the country. With almost 5,000 seats, its interior featured marble, murals, gold leaf, tapestries, and chandeliers. There were three lobbies, a Wurlitzer organ, and the largest chandelier in Philadelphia. Frequently closed during the Depression, it had reopened in

September 1942 with *Tales of Manhattan*, and in the war years it offered premieres of Irving Berlin's *This Is the Army*, as well as stage shows featuring Eddie Fisher, Dean Martin, Jerry Lewis, and Judy Garland.

In those days, movie theaters offered a special "extra." Films were preceded by Movietone footage showing the news. For a world living with news coming into their homes mostly through the radio, it was thrilling and sometimes disturbing to watch the black-and-white clips, sometimes shot up close to the front lines, which told the story of the war as it unfolded.

The other great advantage of theaters was that they often had air-conditioning. To sit in the dark rooms in front of enormous screens and cool air was a relaxing way to spend a few hours after the noisy, intense work of computing.

Late on some Saturday nights, Florence would circulate song sheets with satirical lyrics she had written to popular songs. "[I've Got Spurs] that Jingle, Jangle, Jingle" lyrics became "I've got differences that wiggle waggle wiggle / As I go smoothing crazily along / I've got factors that make me wanta giggle / As I inverse interpolate along / Oh, Gregory Newton, Oh Gregory Newton / How we love your coefficients—Yes, we love them for computin' / I've got ranges that wiggle waggle wiggle / And my 0s waver crazily along / But if Hitler's nerves begin to jiggle / Then our tables can't be very far from wrong."

"Night and Day" lyrics became "Night and Day, all I can see / Are those marks on that long-winded trajectory... / With its X's and X Prime / Oh I make mistakes most all of the time / Whether it's Night or Day!"[30]

On Sundays, their day off, the women went on picnics, trying to unwind from the workweek. "We didn't worry about the men in our life because there weren't any," Marlyn said. "And we all... let our hair down. We did whatever we felt like doing that we could afford to do."[31]

The Monster in the Basement

Like Marlyn and Ruth, Kay and Fran started off calculating ballistics trajectories on desktop calculators with Lila Todd in the early days of the Army computing group when it was still quite small. But their paths quickly diverged from other Computers. Early in their work on the Army computing team, Kay remembers that she and Fran were asked "if we'd be interested in learning how to operate this differential analyzer, which we had never seen."[1] In the basement of the Moore School sat a huge machine. "It was a monster, because it was about thirty feet long and it was all made up of metal shafts and gears that would turn." Someone told them that the Army expected that this machine could solve a trajectory in less than an hour and greatly reduce the time needed to generate a firing table. Kay was a little skeptical because "it was very, very complicated looking."[2]

It was called a *differential analyzer*, the brainchild of Vannevar Bush, a famous scientist from MIT who had created an "analog computer," one that solved sophisticated equations and even some types of differential calculus equations with a huge "collection of shafts, gears, and wires."[3] They were very expensive, and everyone wanted one. In the mid-1930s, when the Moore School was ready to buy, the price tag of $100,000 ($1.5 million today) was too high. So the Moore School approached the BRL to help sponsor the machine. BRL did so with one particular provision: "in case of a national emergency Aberdeen [Proving Ground] would take over the machine."[4]

The analyzer served as a research machine for Professor Cornelius Weygandt and his students until the United States entered WWII. In June 1942, a team from BRL, which also had its own, albeit smaller, analyzer, came to Penn to run the one in the basement of the Moore School.[5] The huge machine was simply too big to move, and the contract gave the Army control of not only the machine but also the room itself. BRL posted a sign on the door that read RESTRICTED. From now on, the analyzer would be used only for BRL purposes. Academic research would have to wait.

To the surprise of the Moore School, which did not have any women working on the analyzer, BRL quickly assigned Computers to run it. Kay and Fran were part of this early team, and on the day they were assigned, they disappeared into the basement to begin their next assignment.[6]

In the former boiler room, they saw something astounding. The analyzer resembled "a giant's mattress spring," according to Professor Joseph Weizenbaum of MIT, who used one as a student.[7] It was over thirty feet and consisted of long metal rods with strings, bands, plates, gears, shafts, and rods. To solve just one ballistics trajectory, Weygandt and his crew had set up dozens of motors, thousands of relays, 2,000 vacuum tubes, and 200 miles of wire.

The analyzer was supposed to be the answer to BRL's need for faster calculation, and indeed when the analyzer was working, it could run a trajectory in less than an hour, far faster than the Computers working on the desktop calculators. To do it, one Computer input data into the equation using a long, flat board. Another Computer watched over the middle of the analyzer, where gears performed multiplication by two when a thirty-tooth gear meshed into a sixty-tooth gear.[8]

Initially, Kay and Fran started on the same shift and it was a joy to be together, as always. They studied under the BRL team, but soon they were separated, as BRL staff, looking to return home, assigned the two young women to be supervisors of the two alternating analyzer shifts: day and night. One would work two weeks as supervisor of the day shift, then two weeks on nights; the other worked two weeks on nights and then two weeks on days. They would miss working together.

Unlike the Computers on the Third-Floor Computing Team at 3436 Walnut Street, where one Computer calculated a single trajectory from start to finish on her desktop calculator, the analyzer teams shared the same equation because it was literally built onto the huge analog machine. Kay's shift would hand off their calculations, and partial results, to Fran's shift, and vice versa. Fortunately, after years of friendship and professional collaboration, the two women could practically read each other's minds. They were the perfect set of supervisors to work together to keep the analyzer calculations on track, sixteen hours a day, six days a week.

Kay, as supervisor, often sat at the opposite end of the analyzer watching as the ten round counters turned to show the results of the analyzer's calculation. When they stopped, she wrote down the results.[9]

But Kay had very little confidence in this expensive and prestigious device and found that "there were millions of things that could go wrong with the machine." To check the analyzer's accuracy, she kept an old and reliable desktop calculator on a table in the back of the room and regularly ran "hand-calculated trajectories just to check on it and make sure that we're not too far off" and check that the analyzer was doing what it should.[10]

When the result was too far off, she called in the maintenance team because the Moore School staff was supposed to fix it. It was their job to build each equation onto the analyzer and then check bands, shafts, gears, and motors to see that everything was running well. A loose band, slipped gear, or broken wire meant that the analyzer's accuracy went down and the trajectory results were less useful. Joe Chapline, who came from Ursinus with John Mauchly to help with wartime projects, had signed on as a maintenance engineer, or as he called it, "the nurse who cared for the operation of the big mechanical computer."[11] Other Moore School students and recent graduates came in and out to help.

The Computers were not supposed to fix the hardware, but late at night when a problem popped up and no maintenance engineers were in the building, Kay sometimes fixed things herself, as she did not want the team to lose valuable calculating time. When the bands were out around the motors, Kay put them back on. When the strings broke, she replaced them with the fishing wire used for the analyzer, a brand called "cutty sark." Years later she shared that whenever she saw a bottle of Scotch whisky called Cutty Sark, she remembered her days working with the analyzer.[12]

But replacing wire under tension was a dangerous task, as were several other aspects of running the analyzer. One treacherous day, when a Computer had her nail ripped out by the analyzer, John Holberton carried her to the hospital himself for treatment.[13]

Despite the sign on the door that read RESTRICTED and indicated that only people with military clearance for the analyzer room should enter, many people found their way into the room. For the analyzer room was air-conditioned, an expensive luxury that was installed—not for the staff, but for the hardware. Its gears, bands, wires, and plates could operate only in a narrow range of temperatures and humidity.

Kay was surprised at how many members of the Moore School violated the military restrictions just to enjoy the cool air. "All the professors at Moore School, plus any graduate students or anybody that possibly found some reason to come to the analyzer room" came down to the remote basement room for their "gab fests," she recalled.[14]

But it wasn't enough. Even with the desktop computing teams and the analyzer computing teams working hard, Herman still could not generate the number of trajectories he needed for the firing tables the Army needed.

By early 1943, he was tearing his hair out.

One day in March 1943, watching Herman standing over the great, analog electromechanical machine in utter frustration, Joe Chapline quietly approached him.

"You ought to talk to a guy upstairs named Mauchly," Joe said. "He has some ideas about how to do it electronically."

"Can I meet him?" Herman asked.

"Sure, come on," Joe said.[15]

Together they went upstairs to meet Joe's friend Dr. John Mauchly. And history was about to be made.

The Lost Memo

Once they met, Herman and John realized they had much in common. Both were teachers, displaced by the war from faculty jobs they treasured, Herman at the University of Michigan and John at Ursinus College. They were both married to mathematicians who had jumped in to help and even run parts of their military projects.

And both were enormously frustrated. Herman because he was under such pressure to produce firing tables and no amount of recruiting, training, desktop calculators, or even use of the differential analyzer seemed to make much of a difference in the backlog. John because, despite his heavy teaching load and helping run the Army radar project, he felt unacknowledged by his peers. He was never accorded the title of "professor," only "instructor" at the Moore School, despite being a well-respected professor at Ursinus and chairman of its physics department.

Plus, John had a vision for an all-electronic programmable computer and was finding it hard to get Moore School faculty to talk with him about it. The computer he envisioned would work at lightning-fast speeds, the speed of electrons, not the turtle's pace of electromechanical switches. It would be digital, not analog, and general purpose, meaning it could solve a wide array of equations and problems.

Herman was intrigued by John's idea about a new computer. The two discussed how this computer might solve ballistics trajectories in minutes, not days, as the hand calculations with desktop calculators could, or hours, as trajectories on a well-working analyzer could.

As John conceived of it, twenty to thirty specially built electronic devices "would be able to perform 1,000 multiplications per second, and to perform complete trajectories in a minute or two."[1] John hoped to use this power for his favorite problem: weather forecasting. He wanted to use computers to calculate the path of storms and provide early warning for communities in the path of big storms (a complex problem calculated by supercomputers today). But his computer would work just as well for ballistics trajectory calculations.

Herman recognized a solution, albeit a far-flung one, when he saw it.

John warned that the price tag for the experimental new equipment would be high, but Herman was not deterred. He knew in the midst of the war, money was not a barrier when important problems were involved.[2] John was delighted that Herman was interested in his vision and dream. Herman's next question flowed easily. Could John write up the idea?

John responded, "I already have."[3]

Seven months earlier, others at the Moore School had asked the same question, and John put time and thought into writing a clear, succinct five-page paper called "The Use of High-Speed Vacuum Tube Devices for Calculating."[4] He then went to a secretary to have it specially prepared, typing on mechanical typewriters being a special skill of the day.

Because copy machines did not yet exist, and carbon paper (a page of ink that transferred impressions of a page being written or typed to the page beneath) was in short supply due to the war, John had presented the only copy of his paper to Grist, who promptly lost it.[5]

When Herman appeared at Grist's office door looking for the

paper, the dean could not find it. Plus, it was clear Grist was not very excited by it. If the Moore School had an analyzer, why did it need an all-electronic computer?

Herman and John were beside themselves for a few days wondering what to do next. John knew his paper would be painstaking to rewrite, but Herman did not want to lose time; he wanted to get John's ideas to BRL as quickly as possible.

A day or two later, John realized that someone might have notes of his paper. Dorothy Kinsey Shisler, a 1941 Ursinus graduate[6] who lived in Philadelphia, had typed John's paper originally. She wrote down John's words in shorthand, a stylized set of abbreviations and symbols for rapidly writing down a person's words, and then typed up the paper.[7] Could Dorothy have kept her notes?

She did! Dorothy had her stenographer's notebook of shorthand with John's paper and felt she could easily retype it. Soon the five-page paper was ready, neatly typed again and laying out a computer unlike any known at the time. It would be all-electronic, general-purpose, and programmable. Perhaps the greatest paper of the twentieth century had just been saved by a woman with impeccable secretarial skills.

The paper shared John's vision of a digital computer capable of performing a broad range of useful work and problem-solving. It would not have to be rebuilt for each new problem but could be programmed and reprogrammed using the same hardware in new and powerful ways. The computer's speeds, if achieved, would be "several orders of magnitude (essentially 1,000 times) faster" than any machine on Earth. The hardware would serve as a platform for testing groundbreaking new concepts of high-speed computing.[8]

When Herman read the paper, he understood the concepts that

John was setting out, including replacing the gears and wheels with electronic counters. He believed John's idea that great speed could be achieved by using vacuum tubes, a key electronic component of the day, instead of gears and electromechanical switches. Herman quickly got interest, but BRL associate director Dr. L. S. Dederick was skeptical, saying that the short paper was not enough and BRL would need a more formal presentation.[9]

John told Herman that he needed someone to help him write the presentation and build the new computer. The person he had in mind was a young engineer, then only twenty-three years old, named J. Presper Eckert who he had met through an electronics course he took at the Moore School. Known to all as "Pres," the young man was a recent graduate of the Moore School and already the holder of several patents.

Herman said yes, but only if the two men could transform the paper into a formal proposal with the steps and costs the project would take. John and Pres "worked 'on a twenty-four-hour basis' for several days and came up with a proposal that contained more details."[10]

Pres was ten years younger, but John recognized a genius when he saw one. The two men both were inventors from birth. In John's case, he was playing with electricity and made money as early as elementary school by installing electric doorbells for his neighbors. John was a "second-generation scientist" whose father was a physicist and worked for the Carnegie Institution of Washington. He loved to read at night, a habit his parents disliked, and according to Joel Shurkin, John arranged a switch on the stairs to trigger the light in his attic room to fool his parents:

> He rigged a trigger on the stairs so that when his mother walked up to make sure he was sleeping, his reading light was automatically turned off. When she went back downstairs, and stepped on the same stair, it turned the attic reading light back on and John went back to reading.[11]

At Ursinus, John became intrigued by the idea of using statistics to see if there were ways in which weather patterns could be more accurately predicted. During summers, he took jobs at the National Bureau of Standards in Washington, where he had access to governmental weather data. He brought the data back to Ursinus and paid students to run statistical calculations of it.[12] But it quickly seemed to John that there should be better ways to calculate these important equations, beyond the time-intensive work of desktop calculations composed of typing numbers, calculating equations, writing results, and retyping numbers, with the myriad errors that such repetitive efforts could introduce.

In 1939, John went to the World's Fair in Queens, New York, and there saw an IBM electronic cryptographic machine that sent coded messages using vacuum tube circuits.[13] Vacuum tubes were glass tubes with their gas and air removed. Within these tubes was placed an electron-emitting cathode and anode that forced the electrons to flow in only one direction, creating a current. Invented by John Ambrose Fleming in 1904, by the time of the 1939 World's Fair, vacuum tubes were in mass production.[14] While they could vary in size, many looked like elongated light bulbs, about five inches long and two inches wide, and could be found in the big wooden radios of nearly every family's living room, there to control the volume of a show by turning the electrical current up or down.

John wondered if vacuum tubes could be used for more, perhaps mathematical calculations by the thousands at the speed of light.

John built his own electronic counting devices using neon tubes, and then realized that what he really wanted was an electronic computer. In 1940, he drove to Dartmouth for a meeting of the American Mathematical Society, where he met an MIT mathematician named Norbert Wiener and they agreed that electronic computers were "the way to go."[15]

But the knowledge of the day, the proverbial "group wisdom," was that an all-electronic computer was impossible and unnecessary.[16] It would require far too many vacuum tubes to operate for longer than a few seconds, and everyone knew that vacuum tubes were unreliable. A full-size computer based on vacuum tubes would fail, according to the leaders of the time.

One of the reasons John loved working with Pres was that Pres did not know such "prevailing wisdom," or was too young to care. Like John, Pres was an inventor almost from birth. At age five, he drew a detailed picture showing his living room radio with its dials, speaker, batteries, and other parts. At age seven, he built his own radio around a pencil that he connected to the metal legs of his desk at school and listened whenever he was bored.[17]

By the time he was in his early teens, he built amplifiers for his phonograph and played his records at whatever volume he chose. Because his father had become a wealthy businessman in Philadelphia, Pres had some unusual opportunities growing up. His house was a few doors down from the famed baseball manager Connie Mack, and he met one of the greatest baseball players of all time, Ty Cobb, with whom his father played golf. On one of his father's business trips, they visited the movie sets of Douglas Fairbanks and Charlie Chaplin—an experience his father wanted to share.[18]

But Pres's greatest dream—to attend MIT—was not one he would achieve. As Pres was the only child, his mother could not bear the thought of her son leaving home and persuaded his father to tell Pres that they could not afford the tuition of MIT and cost of living in Cambridge, Massachusetts.[19] Instead, his father enrolled Pres at the Wharton School at Penn. Bored, Pres quickly transferred to the Moore School. When he found out what his mother had done, Pres was furious.

Nevertheless, engineering intrigued him and Moore School professor Carl Chambers invited him onto several outside consulting projects. Together they worked with Philadelphia-based radio manufacturers and a company working on the new television technology. As a student working part-time, Pres designed circuits and learned that it was worth spending a lot of time in preparation and making sure things were right, both in calculations and in tests.[20]

When Pres graduated from the Moore School, he had several engineering patents to his name and excellent job offers with Bell Labs and RCA. But Pres decided to stay at the Moore School as a graduate student. It was probably the best decision of his life.

In mid-1941, before Pearl Harbor, Pres taught a course for the Army and Moore School called Defense Training in Electronics. The Army knew that it needed more men with electronics aptitude, and the ten-week course was designed to teach modern electronics to men with backgrounds in physics and mathematics.[21] John had enrolled because he hoped he might learn something more about modern computing devices but was disappointed to find that he had already taught a similar course at Ursinus. There was nothing new to learn here.

But there was a hidden benefit of taking the course that John had not anticipated. Pres was the lab instructor, and when the electronics students did not need him, he and John "would sit around on the lab table dangling our legs and spending the hours talking about whatever we were interested in talking about."[22]

More and more they began to discuss all-electronic, general-purpose, programmable computers. John shared his vision and also the prevailing doubts of the day that thousands of vacuum tubes would be needed, but that thousands of vacuum tubes could never work together and reliably to run such a large machine.

But Pres felt differently. He had seen a big organ at a large church in Philadelphia; it had over a hundred vacuum tubes, two per key, and "worked well enough."[23] Pres shared with John that "another factor of twenty was in the cards," or put another way, he could make 2,000 or more vacuum tubes work together.[24]

John felt reassured when Pres saw no barriers to the feasibility of the huge computer they were talking about. The key would be running the vacuum tubes at less than full power, 10 or 20 percent below regular current, Pres determined.[25] The assessment gave John confidence. Without Pres, "I probably wouldn't have been encouraged to proceed," John confessed later.[26] When the labs finished and other students left, John and Pres continued to talk. Together they discussed the design of John's all-electronic computer. For the next year and a half, they met where they could: in the cool air-conditioning of the analyzer room, in various open spots of the crowded Moore School, and at the nearby twenty-four-hour restaurant called Linton's where John drank coffee and Pres sipped ice cream sodas.[27] If you had been in Linton's in early fall 1941, you might have seen two men, one in his early thirties and one in his early twenties, with their heads together sketching inventions on napkins. Neither you, nor they, would have any idea they would become two of the most important inventors of the twentieth century.

"Give Goldstine the Money"

Herman was likely a little skeptical about bringing a twenty-three-year-old recent graduate onto the new project as a lead engineer, but what he wanted most was a proposal, and quickly. If the two men wanted to write it together, and run the project together, Herman would see how it went.

As the two men worked around the clock for a few days, Herman went off to talk with Grist Brainerd. After all, the Army computing group was located at the Moore School, along with the radar project. Why couldn't the new machine be built at the Moore School too? It would need engineers to build it, and many more were available at the Moore School than in the swamps of rural Maryland by Aberdeen Proving Ground.

But Grist was a tough sell. He was largely insensitive to the value of this idea and initially did not want the Moore School involved.[1] He eventually agreed with the other deans that there was not too much for the Moore School to lose. The project would be the Army's, and the fault when it likely failed would be BRL's and not theirs. In the meantime, the Army would pay the Moore School for the space and would help run the project. After discussion, Grist delivered a yes to Herman. It was different when Herman asked than when John asked.

Herman, for his part, felt confident that he could send the proposal to his superior officer, Paul Gillon, and receive a good response. And with Paul's help, Major General Veblen "would be a cinch [because] Veblen believed in people rather than in projects, and he had a lot of confidence that we would do it."[2]

Together, Herman, John, Pres, and Grist prepared the next phase: a formal proposal and presentation, from the Moore School to BRL. Grist wrote the introduction and general portions, analogizing the new computer to a differential analyzer, since that was a machine everyone knew, and John and Pres wrote the technical portions of the presentation, laying out the new computer.[3]

They worked quickly, and Herman and Grist submitted the proposal to Leslie and Paul on April 2, 1943.[4] The paper got their immediate attention.

After the proposal was submitted, Herman and Grist were called to a meeting in Veblen's office. Although Veblen was a Princeton professor and a member of the university's Institute for Advanced Study, along with Albert Einstein, he was still a senior founder of BRL. His opinion mattered greatly, especially on matters involving large expenditures.

In Veblen's office at BRL, Herman explained the ideas for the computer. Veblen sat thinking, his feet on the big wooden desk and his big wooden chair tilted far back. He was in deep concentration. A bit of a debate ensued before him as some officers worried about the hefty price tag of $200,000 ($3 million today) and others worried about the low likelihood of success. The debate went back and forth, and finally, Veblen's chair came crashing down.

"Simon, give Goldstine the money!" and Veblen left the room.[5] But the deed was not done, and BRL needed a more formal presentation. It was set for April 9, 1943. It was time for the full group to go from the Moore School to Aberdeen Proving Ground. Fortunately,

Herman had a Studebaker car and special gas rations for his trips to Aberdeen.[6] As he picked up John, then Pres, then Grist from their homes early that morning, Herman was nervous. Although the project had Veblen's support, he knew that leading scientists and technologists had told BRL that the plan was folly; no one would be able to get a machine with 10,000 vacuum tubes to work.

Would the explanations of John and Pres be able to carry the day?

After about an hour and a half on bumpy local roads, the men made their way from Philadelphia to Aberdeen Proving Ground on the Chesapeake Bay. They passed first through the security of the base and then through the second security layer of the Ballistic Research Laboratory. Herman had the right ID and passes; he had been there many times.

In BRL, the men split up. John and Pres went into a side room to work with a secretary on tweaks to the proposals.[7] Herman and Grist "went in to meet the officers in charge."[8] After Herman and Grist made their proposal, again, Grist was asked to leave. Soon Veblen came out. BRL's answer had not changed; the Army would fund the building of the world's first all-electronic, general-purpose, programmable computer at the Moore School. Colonel Simon shared the good news directly with John and Pres.[9]

The officers went out to lunch with Herman and Grist at the stunning Officers' Club at APG, huge picture windows overlooking the waters of the Chesapeake Bay. But they left John and Pres behind to finish the contract details. The two men felt left out, and hungry.

But they did their work, and when it was time to leave, they took the train back to Philadelphia with Grist because Herman had to stay and finish things.[10] They were happy with the news, and John and Pres were now quite hungry.

It was also Pres's twenty-fourth birthday.

Dark Days of the War

It took no time for the women in the analyzer room to become aware that something unusual was being built on the first floor. They heard rumors of "Project X" but did not know the details.[1]

Kay heard "there was a new calculator being built at the Moore School... but we had nothing to do with it."[2] She could not go into the big room at the back of the first floor of the Moore School dedicated for the new machine because it was classified as "confidential." Its signs indicated that no one without clearance was allowed in the room, and unlike the analyzer room, the restrictions were enforced.[3] Kay shrugged and did not feel slighted. It was wartime, and lots of doors were closed.

As Project X was kicking off, on May 31, Paul chaired a meeting at the Moore School with Herman, John, Pres, Grist, and others in attendance. He had come up from BRL with astronomer Leland Cunningham, a member of the BRL Scientific Advisory Committee, to meet the initial team.[4]

Paul outlined the roles for the new Army contract. Grist would be the project supervisor but would not play a technical role. Herman would be technical liaison to Aberdeen. John, the visionary, would hold the title of "principal consultant," and Pres would be the chief engineer. John and Pres would be co-leads of the design, construction, and testing of the new computer.[5]

In addition to other administrative matters, Paul announced the name of the new machine. While some at the Moore School wanted to call it the "Electronic Numerical Integrator" to analogize it to the analyzer, Paul announced the name as Electronic Numerical Integrator and Computer, and ENIAC (short *e*) was born.[6] It would not be a calculator or integrator as people knew them, but a new machine, a "computer."

Paul shared key provisions of the contract. At the end of construction, if the ENIAC worked, the Army would own it and relocate it to Aberdeen Proving Ground. The contract also gave John and Pres the right to patent their invention and commercialize it later.

This was standard Army contract language for the time and remains similar today. While the Army wanted full use and ownership of the computer it would pay for, Army officers also wanted John and Pres to protect the Army's investment and the two men's invention by seeking appropriate patents and protections for new ideas. The Army also hoped that its research and development contracts might open important inventions to the public. Perhaps this contract might create a new industry?

The agreement was distributed for review on May 17, 1943, and signed on June 5.[7] John and Pres began to assemble their team, and by early July they had about twelve people—"an odd yet powerful collection of engineering talent." In his book *ENIAC: The Triumphs and Tragedies of the World's First Computer*, Scott McCartney wrote that the group included "Bob Shaw, an engineer beloved for his practical jokes and inventive design;...Kite Sharpless, a logical, intelligent, and considerate engineer from a Philadelphia Quaker family; Chuan Chu, a...

Mandarin Chinese immigrant; and Jack Davis, a happy-go-lucky bachelor who speculated in grain futures with Bob Shaw just for fun."[8]

These young engineers were joined by Arthur Burks, a PhD philosopher and logician, who, like John Mauchly, came to the Moore School for the Army electronics class in 1941 and stayed.

John and Pres would set out the overall vision and design principles for ENIAC, but they needed help to design its individual units. Each man was given a specific piece of the machine to engineer, and they disappeared behind the closed doors of the confidential room on the first floor to do so.

In the meantime, the war was not going well and seemed to be getting closer to the United States. The long Battle of Stalingrad, ending in February 1943, was won at a horrendous cost. About two million lives were lost, including men, women, and children, who died of disease and starvation.

The news from the Pacific was also hard, as February brought headlines of a battle on a Pacific island called Guadalcanal. A brutal loss of life occurred for both US and Japanese soldiers, although the event was ultimately a US victory and a turning point in the war. After destroying US ships and troops in Pearl Harbor, the Japanese had now been pushed back and future battles in the Pacific would head in the direction of the Japanese mainland.

There was a bit of good news for the Third-Floor Computing Team. On May 7, 1943, the Germans finally surrendered in North Africa. Since Marlyn and her team had been under special pressure to calculate trajectories rapidly for the artillery in North Africa, they felt a bit of a connection to this victory and a surge of pride in their work.[9]

Despite the positive advances of the Allies in the Pacific, the

Atlantic was a different story in 1943. There, the war seemed very close to home. Philadelphia was on high alert, as it was in range for the feared German submarines, the U-boats. Now, every night, Philadelphia went into blackout mode, and when Marlyn went home after work at 1:00 a.m., it was through darkened streets on a trolley with muted lights.

A year earlier, the U-boats had come across the Atlantic and begun sinking US ships off the coast of New Jersey and Virginia. Philadelphia was between the two and very much in range for the torpedoes that could attack not only ships, but also coastal populations.

Philadelphia city inhabitants, along with residents up and down the East Coast, received directions to buy "blackout blinds" to shield the light coming out of their homes. Other regulations prohibited stores and businesses from displaying their lights to the street at night. A curfew kept people home, so the only people on the streets at night were wartime workers toiling on projects that sometimes ran late or in factories that often ran around the clock.

It was scary for Marlyn to make the trek home through the dark streets of Philadelphia. When she got off the trolley, it was so dark she was sometimes unable to see her hand in front of her face.[10] Nor could she see other people on the street, although sometimes she could hear the sounds of shoes clicking on the pavement and approaching her. She would stand still, hold her breath, and wait until it was more clear who was walking toward her. Almost always, it was the neighborhood air raid warden, the person assigned to check the neighborhood at night for stray lights and broken blackout blinds. He was an old friend of Marlyn's family and knew Marlyn's work schedule well. When he could, he would meet her trolley and walk her home through the quiet and dark streets of Northeast Philadelphia.[11]

Marlyn let out her breath and walked home with her friend.

"All That Machinery Just to Do One Little Thing Like That"

The room for Project X, or "PX" as everyone called it, was a big lab in the back left corner of the first floor of the Moore School. Unlike other security conditions in the Moore School that were sometimes a little lax, the restrictions to this room were enforced and no one not affiliated with the new machine went in or out.

People watching outside could see enormous amounts of supplies begin to disappear behind the closed doors, including large pieces of black metal, thousands of switches, tens of thousands of vacuum tubes, and miles of wire among them.

By February 1944, the ENIAC team finished their wiring diagrams and thoughts turned to building the new computer.[1] John and Pres led meetings on Saturday mornings with their teams to work out the kinks.[2] They had a collaborative style of meetings; anyone could speak and share ideas. "Anybody could express any idea and this would be discussed reasonably," team member Arthur Burks remembered.[3]

The hallways of the Moore School filled with new faces as newcomers, men and women, found their way to the PX room and disappeared behind the closed door. Kay noticed the new employees showing up, and eventually a group of women was hired to work on wiring.[4] Herman thought to hire women who had telephone wiring experience:

> Moonlighting telephone company workers, who didn't have much regular work since the war was on and few companies were opening new offices, were hired to do much of the wiring inside the machine.[5]

These temporary employees came and went, but the two men supervising them, Joseph Chedeker and his assistant Sol Rosenthal, a concert cellist, stayed on.

Gradually, the units of the new computer began to rise.

The ENIAC was of course only one of many unprecedented and top-secret innovative activities developed by the US military during the war. Betty was dispatched briefly to another computing project at Pine Camp in upstate New York, about five hours north of Philadelphia, and she was very cold. Now called Fort Drum, Pine Camp was an Army base just south of Canada, used for winter military training and founded in 1908.[6]

Betty went to Pine Camp "for a session" and did some computing work. On a desktop calculator, she ran calculations for weather forecasting, based on weather balloon experiments taking place at Pine Camp.[7]

But these were no ordinary weather balloons, and this was no ordinary Army base. Pine Camp was part of an elaborate deception created by the Allied forces under General Dwight D. Eisenhower to hide when and where millions of Allied troops from the United Kingdom would invade Continental Europe.

In one of the greatest disinformation campaigns of all time, General Eisenhower ordered General Patton to pretend that he was taking a great force over from the United Kingdom to land in France at the town of Calais, the closest French town to the United Kingdom and the shortest ride across the choppy English Channel. The goal was for the Germans to locate their main defense forces in Calais, along with the biggest artillery guns and main fortresses.

Fake messages were sent for the sole purpose of being intercepted and read with information about the Calais landing. An entire fake "ghost army" was created to give life to the misleading information. The soldiers "used inflatable tanks and artillery, sent false radio transmissions, and blasted audio recordings of troop movement and construction to create phantom forces" to mislead German forces about the location and size of US troops.[8]

Much of the audio recording technologies and techniques came from Army forces working at Pine Camp. They worked with employees of the nearby Bell Labs research facility who came to assist with the work. In the end, the soldiers could "produce soundtracks [of troop movement] that were impossible to distinguish from the real thing."[9]

It was a new method of warfare called *sonic deception*, and it soon moved from Pine Camp to the European war theater. The soldiers from Pine Camp were sworn to secrecy about the project for decades.

Years later, Betty learned that her group was part of the deception. Another group of Computers in the capital were calculating the actual weather equations for General Eisenhower's D-Day landing, and the Army wanted a diversion. "We were sort of camouflaging the weather forecasting that was being done in Washington."[10]

When her weeks at Pine Camp were over, Betty was happy to head south back to Philadelphia. For when the wind blew on the mountaintop where Pine Camp was located, Betty had never felt colder.

Meanwhile in the analyzer room in the basement of the Moore School, Kay's team of specially trained Computers was regularly being raided. Those frequent visitors to the analyzer room did more than talk and cool down. On lunch breaks and after shifts, they began to date the women on the analyzer teams, and then more. "I had a wonderful girl working with me by the name of Alice Rowe," Kay shared, "that Dr. Burks came down and fell in love with, and took her off." She had another Computer, named Marjory, and someone else from the Moore School "came down, married her and took her off."[11]

Computers did not have to leave their positions just because they got married. Kay worked with Alice Snyder, married to a doctor at the University of Pennsylvania Hospital, and Sis Stump, whose husband had been "drafted into the Army, so she had taken a job there at Penn during the course of the war."[12] But it was a choice women had, and many chose to take it.

This left Kay, with a sigh, often retraining new Computers to run the difficult-to-manage analyzer in the midst of the war.

Many times when she had a free night after working the day shift, Kay visited wounded men at veterans hospitals or attended USO (United Service Organizations) dances. Although her work left her with very little free time, Kay felt it was the right thing to do: "We always felt that those soldiers were doing an awful lot more than we were doing."[13]

If she was going straight from the analyzer room to the dance, Kay went to a file cabinet in the corner, took out a pair of heels that she kept inside, and slipped them on before she clicked out, up the stairs, for an evening of fox-trotting and swing dancing.

On long nights, when the analyzer was running well and could be run by two people, Kay and the others read books to each other. Kay read aloud *Look Homeward, Angel*, Thomas Wolfe's first novel about a young man who leaves home at the age of nineteen, to her team.[14] It helped to pass the time.

On one such night, John Mauchly and Pres Eckert "came bouncing down to the analyzer room." Kay was among those who recognized them from their time hanging out in the analyzer room for their "gab fests." The men said, "Do you want to come and see what we've accomplished?"[15]

The women looked up, a little dazed. They were deep in concentration in the quiet of the late night at the Moore School. It was strange to be interrupted in the middle of their project, and the two other women looked to Kay for guidance, as she was their supervisor.

The two men were so excited that it was hard to say no, and Kay and her team took a break, carefully locking the door behind them. They walked behind John and Pres, who practically flew up the stairs. The men were bursting to share some good news.

"They took us to an enclosed area in front of the room where the PX was being built," Kay recalled, "and near an enclosed area about 8 feet square with a big sign that said, 'Keep Out High Voltage.' "[16]

Inside the enclosed area, a big metal cage, stood two big units, each eight feet tall and dwarfing even John, who stood about six feet tall. The units were approximately two feet wide and encased in black metal except for places on the front with switches, wires, and a grid of many small lights at the top.

"Look at that," John and Pres said excitedly.[17]

The two units were attached to each other with thick black wires, and one of the men pushed a button. Suddenly, and for only a second, the lights on both units twinkled at a lightning pace, turning on and off in an unusual pattern. Then they stopped and tiny lights at the top of one of the two units shone brightly.

"Look at what?" the women responded.

The two men explained that the women were looking at two ENIAC accumulators, and the two units had just added five to itself five thousand times.

"How do we know?" Kay asked.

John and Pres answered with huge smiles. The two accumulators had just performed 5,000 additions in only one second![18] The result was 25,000, which they explained was stored in the accumulator with the tiny lights shining.

The two men shared that on this night, the two accumulators had worked for the very first time. Knowing that the first two accumulators worked, John and Pres were pretty certain the rest of ENIAC's design would work. It was a milestone for the two men and "proof of concept" that their ideas for modern computing would work.

The women congratulated them warmly, understanding now they had been in a special place at a special moment. It is an honor to be witness to a new invention, and it is a sign of respect for inventors to ask you to be there. Kay would remember the "two-accumulator test" for the rest of her life.

But as they walked down the stairs by themselves, back to the analyzer room in the basement, the women were a little more skeptical. "[A]ll that machinery just to do one little thing like that," Kay recalled. "We couldn't believe it."[19]

In the days to come, Kay wondered about what she had seen. What would Project X look like when it was finished, and who would use it? Would its users have to be engineers? Would they have to be men?

The Kissing Bridge

In the winter of 1944—with every Computer team at 3436 Walnut Street working seemingly nonstop, six days a week, and Kay and Fran doing everything they could to keep the analyzer accurate and producing trajectories, and the ENIAC still far from completion, and no end in sight for the war—Herman and Adele continued to recruit women math majors. Now they had to look beyond Philadelphia and New York, areas they had already scoured, for women in mathematics across the country.

One thing they did not know was how many women were in college studying mathematics. To find them, Herman and Adele asked a mathematics society to send out their recruiting letter. When Dr. Ruth Lane—at Northwest Missouri State Teachers College (now Northwest Missouri State University)—saw it, she immediately had a student in mind.[1] Jean Bartik, about to graduate, had a special drive and ambition.

Dr. Lane knew Jean was finishing her senior year early, in December 1944, and looking for a special job, one with opportunity and challenge. Dr. Lane, who had worked at the huge Wright-Patterson Air Force Base in southern Ohio and then come to Missouri, knew that the military offered some interesting opportunities and challenges, including some that women at the time could not find in private companies. Lane shared the letter with Jean and asked her if she wanted to apply.[2]

Jean, in turn, shared the letter with her adviser, Dr. Joseph Hake, also head of the math department, and he discouraged her. Jean could "go to a big place like Philadelphia," but she would "just be a little fish in a [big] pond."[3] He urged her to stay close to home where her skills could be put to good use in the local community.

But Jean wanted to try something new, something that no one in her family had done. She did not think she would get lost in a big city, and she was ready for an adventure. One brother was in the Navy and if the Army needed her as a Computer, she would leave Missouri, her family, and family farm and travel 1,100 miles away from everything she knew to go to Philadelphia.[4]

She sent her application to the address in the letter, finished her courses, and completed her requirements for graduation. Then she went home to her family farm to wait for the news from the Army. It took a long time for the answer to come.

Jean grew up on a family farm in Alanthus Grove, Missouri, an intersection of a few buildings and family farms in the northwestern part of the state. She attended the one-room Jennings Schoolhouse with students from the nearby farms, mostly her siblings and cousins. The same teacher worked with one class and then another. Jean "always felt it was an asset" because she could "sit and listen to the recitation of the other classes" when she had finished her own schoolwork.[5]

The sixth of seven children, she always felt she had to catch up with her older brothers and sisters and that it was hard to do something they had not done before. She was very good in mathematics, but "everybody in my family was good at math," she would say, and "this was not extraordinary at all."[6]

Jean did chores on the farm, as her entire family did, and could milk a cow before she ever went to school. With two older sisters, "my mother had the cook and dishwasher," so Jean worked outside with her father and brothers.[7] She plowed the fields with a team of horses in the spring and picked crops in the fall.

After school and chores, she went over to Grandma Jennings's house down the road to gather eggs from the chicken coop. When she was done, her grandmother would give her a thick slice of fresh pound cake and they talked, just the two of them, with her grandmother sharing the family stories.[8] Jean learned of her family's long history in Gentry County, Missouri, as farmers and schoolteachers. Not only had her great-uncles and uncles attended the state college down the road but so had her grandmother and aunts. Education in the Jennings family was for everyone, men and women. Jean never questioned that a good education was something to which she was entitled.

She and her brothers helped her father, and when they finished on their farm, they went over to help Uncle Fred at Grandma Jennings's farm nearby, which he ran. Jean was proud that she could hoe as much corn as her brothers, but she was troubled that her uncle, who did compensate them for the labor, paid Jean only half as much as he paid them. Fifty cents a day to the boys' one dollar.[9] It was an inequity she would remember her entire life, and she made a decision henceforth to fight for fairer pay.

When Jean was ready to go to high school—at the age of twelve, having skipped fifth grade—she had to move to Stanberry, Missouri, because the local high school, where all the small schoolhouses sent their students, did not have buses out to Alanthus Grove, six miles away.[10] Also, later in the school year, the thick Missouri snow would make the local road impassable. It was better to be in town.

Jean moved in with her older sister Erma and cleaned house for Erma and her husband in exchange for room and board. She enjoyed living in a larger town, a group of about 2,000 people rather than the 200 or so residents of Alanthus Grove. For the first time, she lived among town residents who were not family members.

She liked Stanberry High School and dove deeply into its clubs and classes. She joined and eventually became president of the student council and an editor of the high school news section of the *Stanberry Headlight*, the town's weekly newspaper. Jean took every math and science class she could find, including algebra, geometry, trigonometry, and physics.[11]

She played volleyball and other sports, but the sport that made her famous in town was softball. She was starting pitcher for the high school girls' softball team, and "[e]verybody came out to watch the girls play." Their crowds were so big that the "boys' team liked to play doubleheaders with us because we drew a large crowd."[12]

Jean's athletic prowess brought her a good deal of notoriety, a bit to her mother's dismay. One day, when the owner of the pool hall came over to talk with Jean about a recent game, Jean's mother stopped the discussion. She was no fan of bars or pool halls, and she felt it inappropriate for him to approach her high school daughter to discuss the game, but Jean was happy.[13] People knew who she was and were watching her team. When she pitched a no-hitter, it was the talk of the town.

Jean graduated in June 1941, at the age of sixteen, second in her class.[14] Then, following in the footsteps of three generations of her family, she entered Northwest Missouri State Teachers College about twenty miles away. With the many unpaved roads of the area, it was too far to commute, so Jean moved on campus to start her college years.

She moved into Holt House near campus, a residence that rented rooms to twelve female students with a lounge area for the girls to socialize and a kitchen for them to use. She was in the lounge on December 7, 1941, playing bridge with her friends freshman year, when someone ran in and said that Pearl Harbor was being bombed by the Japanese. "Everyone immediately gathered around a radio and listened in horror as the news of the devastation was broadcast from Hawaii."[15]

The next day the women of Holt House, together with the entire campus and country, gathered again around radios as President Roosevelt asked Congress to declare war on Japan. Three days later, on December 11, 1942, Japan's allies, Germany and Italy, declared war on the United States. "We realized our lives had changed forever," Jean recalled.[16]

Spring semester on campus was thus entirely different. About half the campus was gone. Male students had gone off to the military, and many male faculty members too. Suddenly the coed campus felt like a women's college, and Jean was not happy.

She was worried about "friends and relatives who were fighting overseas."[17] To a lesser degree, she was worried because many of the classes she wanted to take, including math and physics, were largely male and might not be offered if there were no male students.

Fortunately, the next semester, the Navy decided to turn the campus into an officer's training program and brought four hundred sailors to campus. Now Jean's classes in analytic geometry, trigonometry, and physics were filled with sailors. She was often the only woman, but it didn't matter. She was excited to have full classes and happy to meet her new classmates. When it came time to pick a partner in physics lab, "a dozen sailors ask[ed] to partner with me."[18] She enjoyed talking with them and made many friends.

Dances came back, as did football and basketball games. The campus was once again filled with music, dancing, and cheering.

Soon, Jean began dating a sailor named Joseph Amad. He had "gorgeous curly, black hair" and was "incredibly sexy." He was also a "thoughtful, modern thinker," she remembered. He added a "dash of excitement and romance" to her life in an otherwise difficult time with news of the war.[19]

One place the two of them liked to hang out was the small, rustic wooden bridge on one side of the campus. It was known to everyone on campus as the "kissing bridge," and Jean and Joe liked spending time there.[20]

But all too soon, less than a year later in December 1943, Joe told Jean that it was time for him to ship out to Albuquerque, New Mexico, over 800 miles away. She cried, but in private, because she did not want to make him feel badly. Although "rushing to the altar" to get married "before a soldier shipped out was quite common in those days," Jean and Joe decided it was not the path for them.[21] The future was unclear for both of them, and they agreed to try to stay in touch.

They hoped for a little privacy at the kissing bridge to say good-bye before Joe left, but it was too cold that December night to spend much time outside. Instead, they said good-bye, passionately, at the train station, not caring who might see or comment. As Joe's train pulled out at midnight, tears fell down Jean's cheeks, and Joe leaned out the window and waved for as long as she was in sight. Jean felt she would never see Joe again.[22] They kept in touch by letter for a while, but she never saw him again.

Without Joe, there was a big hole in Jean's life. She filled it with women friends, basketball and golf intramural teams, calculus and astronomy classes. But spring of her junior year was hard and lonely.

In June of 1944, some good news arrived. The United States had launched its invasion of Europe from the United Kingdom on the beaches of Normandy. Like the rest of the country, Jean read about the surprising landing across miles of Normandy beaches in the local papers and listened to excited reporters on the radio news. In the end, almost three million American, British, and Canadian soldiers would arrive on those shores, establish a beachhead, and proceed east across France toward Germany.

First, however, they had to win the beaches, and German artillery and machine guns were in fortified positions aimed at the beaches along the Normandy coast. Thanks to the deception about where General Patton would be landing, the deadly and maneuverable German Panzers were still centered at Calais and far from the Allied landing site.

The honor of being first to land, after the choppy English Channel crossing, went to the Free French Forces, coming to free their homeland. Millions of men would follow, battling to advance up the shores and cliffs. Entrenched German cannons in fortresses on the cliffs rained machine-gun and artillery fire down on the Allied soldiers who fought for every foot of the beach, scaled the cliffs, and disabled the German artillery at the top. The country, and the campus, celebrated as the US and Allied troops began to move east across France toward Germany, while still worrying about the difficult and deadly battles ahead. Soon there would be a large cemetery maintained by the citizens of Normandy in respect and appreciation for the young men far from home who had come to help them.

In addition to the war worries of the summer of 1944, Jean had financial worries. Her favorite aunt Gretchen, a high school principal in Ohio, had helped her pay for her first two years of college with a loan, and her father had helped to pay for her third year of college (Jean worked too), but she would have to pay for her fourth year of college, he told her at the start of the summer of 1944.

She worked at the Pratt & Whitney plant in Kansas City that summer, silver-plating "a little gear that fit behind the propeller of an airplane." It was rough work, as to "plate the gear, we first had to degrease it by handing it on a cable strung across a big degreasing pit" and then "scour them with steel wool." Then she and her coworkers hung the gears in a silver cyanide plating bath. The fumes of the pit made them dizzy, the steel wool scratched up their hands and nails, and the silver cyanide went into the scratches caused by the steel wool and "caused sores to bubble up on the skin."[23] And they worked from three in the afternoon to eleven at night seven days a week.

It was a miserable experience, but Jean stuck with it until late August. When she left, she and her partner were told "that we had turned out more good gears than anyone else had ever done." It was good news, but it did not change a bad experience. Jean was happy to leave and return to campus.

There, even more challenges awaited. With her savings from the summer, Jean had only enough money for one semester. Plus she needed twenty-two credits to graduate, too many to earn in a single semester, a dean told her. To top it off, two of the math courses she needed to graduate were not offered that semester due to an absence of students, the sailors having left the campus.

Jean went to the head of the math department's office and cried. She never forgot what happened next: "Dr. Hake leaned back and said, 'I am head of the Math Department. We can hardly give a degree in math if we don't even give the courses needed to get one.' "[24]

He asked two retired teachers to offer the two courses and arranged them around Jean's schedule. She took all twenty-two credits in the fall semester and finished her requirements for graduation.

Jean soon began looking for jobs, and that's when Dr. Lane, her former calculus professor, showed her the letter from Aberdeen Proving Ground seeking women math majors. Jean had just become one.

She was home for the early spring, and nearly every day her father, a local schoolteacher as well as a farmer, shared openings for local high schools seeking teachers, including math teachers. It was hard, but Jean kept saying no. She believed the Army would respond to her application . . . someday.

At the end of March, almost three months after her application, a telegram arrived from the Army to Jean. The local Western Union telegraph dispatcher in Stanberry did not know how to reach Jean on the farm, so he delivered it to Erma, and Erma called Jean on the phone to read it.

It was good news! The Army had accepted her as a Computer.[25] As Jean cradled the phone's thick black receiver, she looked out over her family's fields just turning green and the family's windmill that pumped the water for the house and farm animals, and paused. "I had a few moments of desolation at the thought that I would only see my family on visits."[26]

But Jean was not turning back. She had applied for the job, and the Army had accepted. The telegram asked her "to come immediately" to Philadelphia and Jean took that phrase literally.[27] The next train east was the very next night at midnight. It was more than enough time to pack her belongings, but barely enough time to say good-bye to her large and warm family.

Erma lent her $100 to get started, and Jean bought her train ticket.[28] That night, Jean's father brought her to the train station and the two of them had a sad moment of good-byes. Her father's dream was for all of his children to live within ten miles of Alanthus Grove, and one by one they were leaving.

He told her that he would probably never be able to visit her in Philadelphia but that she could come home "anytime I wanted—[and] he would be there."[29]

At midnight, right on schedule, Jean boarded the Wabash Express,[30] a great steam-powered train made famous by its many trips crisscrossing the country and the songs sung about it on guitars and banjos. Like so many others during the war, Jean was off to an entirely new part of the country and to find a new part of herself.

Are You Scared of Electricity?

Jean paid $35 for a ticket to go over a thousand miles east to Philadelphia, and she stayed up all night. She had to change trains in St. Louis the next morning, and then "rode all that day and the next night, across southern Illinois, Indiana, Ohio and the long state of Pennsylvania."[1] She would never have dreamed of splurging for a ticket for a sleeping car, so Jean stayed in the main cars, which were very crowded.[2] As she rode, she thought about Philadelphia and what she had learned from her research in the winter after she applied.

Her new home was the city of Benjamin Franklin and his great inventions, including the Franklin stove, with its enclosed top, bottom, sides, and back that kept the heat from disappearing up a chimney; volunteer fire departments; and public libraries. It was home to the Continental Congress that wrote and adopted the Declaration of Independence, and it housed the Liberty Bell that pealed the Declaration's adoption in 1776. A decade later, Philadelphia would be where the US Constitution was written, in secret, to create a stronger central government among the states, and the United States would come into being. Jean could not wait to see Independence Hall, where so much had happened.

Philadelphia was also the home of Stetson hats and the great Wanamaker's department store. In WWII, it was a city of about 2.5

million people, and that was a little overwhelming. Kansas City was the largest city Jean had been in, although that had a population of only about 400,000 the summer Jean worked there. This was going to be a wholly different experience.

Jean arrived in Philadelphia on March 30, 1945. She disembarked at the North Philadelphia station and took a taxi to downtown Philadelphia. As she looked around at the buildings and people of the big, old city around her, she thought, *Nobody here knows anyone from my family.*[3]

She dropped her bags at a Young Women's Catholic Association (YWCA) downtown, where she could rent a simple room for $2 a night and use the gym facilities. She quickly took another taxi to the University of Pennsylvania, where the telegram had stated she report for work. "Needless to say, they were shocked that I had gotten there so quickly."[4] Although the telegram urged haste, the Army assumed it would be another week or two until she arrived.

Army staff gave Jean her sign-up paper, a "War Department Notification of Personnel Action" dated March 30, 1945. It was addressed "To: Miss Betty J. Jennings," Jean's legal name, "THROUGH: Lt. Landry, Ballistics Research Laboratory Division." Her position was Computer, her rank SP-6, her length of employment, "War Service Indefinite Appointment," and her starting salary was "$2,000 per annum," a little higher than when Kay and Fran entered in 1942.[5] Plus she would earn $400 a year for working overtime on Saturdays.[6] It was a document she treasured.

But first she had to take a physical, and since the doctor recommended by the Army staff was near Penn, Jean went straight there. He saw her, but it was an odd physical, "overly familiar."[7] The doctor said he had to finish the physical over the weekend and invited her to his home, but Jean refused. "[T]he old farm boys had taught me well to stay out of secluded places such as haylofts with them." She came back to his office on Monday to the follow-up appointment that he scheduled, and again the doctor acted inappropriately. Jean reported him immediately. "When I informed the manager of the project what kind of lecher he was using as a doctor, he was no longer recommended."[8]

Jean was assigned to Adele's advanced Computer class, and there, the far reach of Herman and Adele's recruitment efforts showed. There were two women from Kansas, one from Ohio, one from Wisconsin, and, of course, Jean from Missouri.[9] These five women, so far from home, became friends.

Adele impressed Jean the moment she entered the room. With her sophisticated way of dressing and thick Brooklyn accent, she lectured with ease about numerical analysis and inverse interpolation and was an "absolutely, very serious, no nonsense, very good teacher."[10] Jean, never shy, participated actively in the class and asked a lot of questions, which Adele seemed to like. Adele became Jean's new role model.

Jean joined Lila Todd's group, now computing trajectories in the fraternity house, and got to work full-time behind the Monroe calculator to which she was assigned. When she had questions, there were plenty of women who could answer them. Lila Todd and many others had been there for much of the war.

Sundays were glorious, and she spent them "running around seeing the sights of the city" with her new friends from the Midwest.[11] They went to Willow Grove Park, the then-fifty-year-old amusement park, the Philadelphia Zoo, and of course, the Philadelphia Museum of Art.

When it was time to relocate from the short-term housing of the YWCA, Jean was unsure what to do. Philadelphia had a severe housing shortage with the wartime housing boom, and many boardinghouses offered little space or privacy. Some workers shared beds with each other, as those leaving for the day shift ceded their berths to workers just arriving home from the night shift. Jean did not want that arrangement.

Fortunately Penn had a good housing department, and the staff sent her to a beautiful neighborhood three-story house owned by a woman who rented rooms mostly to students of the Curtis Institute of Music.[12] Still only twenty years old, Jean fit right in with the students, and they showed her a whole new world of music. She went to their local concerts nearby and to the great Philadelphia Orchestra performances downtown where she listened to a world-class orchestra, with her landlord's son-in-law as the lead horn player. Jean was thrilled.

On April 12, 1945, just weeks after her arrival, Jean and the rest of the country learned that President Roosevelt had died. The next day, the *Philadelphia Inquirer* proclaimed, ROOSEVELT DEAD: SUCCUMBS TO STROKE AT 63.[13] Vice President Harry Truman had already been sworn in as President by the Chief Justice of the Supreme Court.

April 13 was a Friday, and while everyone went to work, all were working in slow motion. Passersby wept openly. When Jean went to a nearby lunch counter to eat, many people sat at their tables, crying and reading newspapers edged in black.[14]

Jean and most of the other Computers could barely remember another president. President Roosevelt took the oath of office on March 4, 1933, and had been elected four times since. Jean had been only nine years old at his inauguration.

It was overwhelming to lose the leader who had helped turn the country around during the Depression and who had been the voice they listened to and followed during the war. President Roosevelt's "fireside chats" revolutionized the use of radio technology and helped calm the public's fears as the President held thirty-three evening conversations with the American people from 1933 to 1944.[15]

It was the first time a president regularly used the radio to share his messages directly to families. Sitting next to his fireplace, President Roosevelt talked to the American people in their living rooms, sitting by their own radios and fireplaces. In a calm and comfortable tone, he explained the tough issues of the day, initially the jobs, banking, and economic crisis, and later the horrible war they had entered. More than sixty million Americans listened as Roosevelt explained, and no president had ever felt so intimate or connected to their lives.[16]

Afterward, things moved quickly and on May 7, 1945, Germany surrendered. On May 8, Philadelphia and cities and towns across the country celebrated. On May 9, Jean and her team, as well as the country, returned to work. There was still no end in sight on the other front, the war in the Pacific with Japan.

Day in and day out, Jean continued calculating trajectories for the artillery guns now being sent to the islands of the Pacific, as the US soldiers fought difficult battles on island after island toward Japan. As long as the Army needed her, she was there to help, but her daily work with the Monroe calculators was hard, and she felt badly for the women who had been there for several years.

In June, a memo came around to the Computers that would change everything. It said that Aberdeen Proving Ground sought math majors to work on a new machine being built at the Moore School, and "[w]e were invited to interview for the job."[17] Jean was invited to a meeting but was a little disappointed when twelve other Computers also gathered in the meeting room, all with more experience. But she stayed and watched as they were called into a back room, one by one, for individual interviews.

When it was her turn, Jean found Lieutenant Herman Goldstine seated at the table and met Dr. Leland Cunningham, an astronomer and member of the BRL Scientific Advisory Committee. Herman led the questioning, and after some small talk asked, "What do you know about electricity?"

"I've taken a physics course," Jean responded. "I know that E equals IR."

"Well, that's not what I mean," Herman responded emphatically. "My question is, are you afraid of it?"

To which Jean responded with a laugh, that no, she wasn't afraid of electricity.[18]

As they were wrapping up, Adele came into the office, saw Jean, and nodded to Herman before leaving. Jean felt that it "was some kind of signal" to Herman, that Jean, the student with many questions, could handle the new job.

After a few days, Jean learned that five Computers had been selected, plus two alternates, and she was the second alternate. *Well, that's the end of that*, she thought. But one of the Computers did not want to give up her apartment for the summer, a requirement as the work would take place at Aberdeen Proving Ground, and the other already had vacation plans.[19]

On a Friday afternoon, "I was called into the office of Lieutenant [Leonard] Tornheim, the overall manager of the computing group at the fraternity house, and asked if I could be ready to go to Aberdeen on Monday."[20]

There was only one answer, and Jean happily almost yelled "YES!"[21] The job was hers, and once again, with practically no notice, Jean went home to pack for a trip by train to a place she had never been, on a new adventure.

Learning It Her Way

The end of the war in Europe had wide-reaching effects on every American. The spring of 1945 was an exciting and emotional time for so many, including the Computers at the Moore School. For Kay, still leading the long shifts working with the analyzer, Germany's unconditional surrender on May 7, 1945, was a cause for celebration.

"We were all jubilant...they actually gave us the day off."[1] The next morning, May 8, was the official Victory in Europe Day (VE Day), and Kay, Fran, Alice Snyder, Betty, Joe Chapline, and a number of others from the Moore School agreed to meet in town near City Hall. It was a good thing they arranged a place to meet because hundreds of thousands of Philadelphians poured into the streets downtown, cheering, laughing, and dancing.

Kay remembered the scenes as people were "just letting loose, just talking to everybody, hugging everybody, yelling their heads off. And it was just fun."[2]

But the next day, Kay, Fran, Alice, Betty, Joe, and the others went back to work. Kay's brother Pat was still fighting in the Pacific on the ship of US Navy Admiral William Halsey, the commander of the Third Fleet, along with hundreds of thousands of US sailors and soldiers.[3] In February and March of that year, nearly 7,000 soldiers had been killed and 20,000 wounded fighting Japanese troops for the island of Iwo

Jima. Even as Kay and the other Computers were celebrating VE Day, US troops were fighting an even bloodier battle for the island of Okinawa, one that would last into the middle of June 1945. Although US forces steadily were moving closer to Japan, the war in the Pacific was not close to being over, and President Truman and other government officials warned that it might continue for years.

At the same time, ENIAC was nearing completion. Although it was about a year after Herman predicted, and BRL was a little upset, the delay was understandable. Parts had been hard to get, and some, when received, had to be reordered due to problems. Skilled personnel had been hard to get, and John and Pres had kept their teams working day and night, often seven days a week, for what seemed like years. All in all, the delays were understandable, but still, BRL felt a special need for this new computer and a continuing desire to speed up the calculation of trajectories for the new artillery guns ready to be shipped to the Pacific.

As ENIAC neared completion and eventual transfer to the Proving Ground, Herman started to think about a staff that would run and maintain the ENIAC after it moved to BRL. He asked Lieutenant Tornheim to hold a meeting of eight computing supervisors, including Lila Todd.[4] They sat down to brainstorm.

The Army needed five women to go to Aberdeen that summer for about six weeks to learn to operate the IBM card reader, card punch, and tabulating machines (tabulators printed cards onto full-size sheets of white paper, connected accordion-style), the only non-unique units of ENIAC. In the interest of time, John and Pres had agreed they did not need to re-create this equipment that would "input" and "output" data, but would focus on the core processing and programming units of ENIAC, now numbering forty units, each eight feet tall and two feet wide.

The commitment would be great. In addition to spending much of the summer of 1945 in the Proving Ground, the women selected would have to agree to relocate to the base after ENIAC's transfer to continue its operation and to teach others. To staff the group, Herman wanted to assign "six of the best [C]omputers to learn how to program the ENIAC."[5] He just had to find them.

Some of the senior supervisors wanted to volunteer, but Herman urged them not to. He needed them to continue the work of their teams, now proceeding at a regular clip. BRL had hoped to have ENIAC already to do the trajectory calculations, but in the absence of a new computer, the women's teams had no choice but to continue.

He asked each supervisor to select one Computer from her section for the new project, and they agreed to get back to him. Soon he and John Holberton were approaching candidates.

John Holberton asked Marlyn if she would be interested. "John said that they were picking six people to work on a special project and they wondered if I would be interested in doing this," Marlyn recalled, and John noted that while she would no longer be working with her Third-Floor Computing Team, they would be close by.[6]

The icing on the cake was that Marlyn learned John Mauchly was involved in the project. This was fine by Marlyn "because I had worked for him before and I enjoyed it."[7] Marlyn said yes.

Then, not quite following his own rule, Herman began to approach a few select supervisors. He came to Betty one day during her shift at 3436 Walnut Street:

> I happened to be a supervisor of girls doing desk calculation for trajectories and I got awfully tired of that and so one night,

> Lieutenant Goldstine came and talked to us, and he told us he couldn't tell us about the machine, because he said it was a secret project, but he asked us: would we be willing to go on this project?...I said to him, anything is better than what I'm now doing. That's all I ever said. I was never interviewed or anything.[8]

And Betty became a Programmer.

Herman came to Kay on a shift down in the analyzer room. He asked if she would spend six weeks that summer in the Proving Ground where she would learn about the IBM reader and card punch. "I knew as much about IBM equipment as I knew about ENIAC, and that was nothing, but I was willing to go," Kay said, laughing. And when Herman asked if she could move with the new machine to the Proving Ground later, Kay agreed. "I don't see any other prospects at the moment," she said a bit vaguely.[9] It was enough of a commitment for Herman. He had another Programmer.

Ruth found out, as Jean had, from a notice circulating among the computing teams, "a petition to the whole Ballistics Research Lab and they asked if anybody was interested in learning to work on a special machine that was classified, and they couldn't tell anybody about it until they were chosen...About sixty of us signed up because we were all curious and I was one of the lucky six that was chosen."[10]

Ruth became a Programmer.

With Jean moving up as second alternate to fifth slot, that rounded out Herman's team for Aberdeen Proving Ground.

Herman did not share that he had a sixth woman in mind for their team, Fran. She was mathematically and technically minded, but he could not lose both of his long-term analyzer supervisors at the same time. He did not tell the five women that he had arranged to assign Fran later to the team.

On a Monday morning in mid-June 1945, five women met on the platform of the Baltimore and Ohio Railroad (B&O) train station located at 24th Street and the Chestnut Street Bridge. The tracks ran by the Schuylkill River in Philadelphia, past Wilmington, Delaware, alongside an entrance to the Chesapeake Bay, and then south to Aberdeen, Maryland, where they would be disembarking.

Many of the women knew each other: Marlyn and Ruth, of course, and Kay and Betty. Jean was new to everyone, and they welcomed her warmly. They compared notes, and no one knew much about what they would be doing at the Proving Ground, but they would be doing it together and they were glad to be in a group.

The B&O train arrived, and about an hour and a half later, they got off at the Proving Ground and switched to a shuttle of small train cars that moved around the huge base. They had instructions to get off the shuttle at the platform with the headquarters (HQ) building, be processed through HQ, and then walk about a hundred yards to the entrance of the Ballistic Research Laboratory.

They showed their paperwork and got their credentials to come on base, walked to BRL, and then found a small base within the base, along with another security checkpoint. Herman had given them the clearance paperwork, and they entered BRL. Administrative officials showed them their classroom inside a newly built, bright-red, three-story brick building. Then they left BRL to find their dorms on the main part of the base.

Along the way, they saw huge open green spaces with thousands of new soldiers being trained for battle. The population on the base had ballooned during the war to over 32,000 soldiers and civilians working there.[11] Major construction had taken place before the women arrived, including a second hospital barracks, four new wooden chapels, and many other buildings.

The group arrived at their living space, a one-story women's dorm, more like barracks than a hotel, at the far end of the base. It was small because only a handful of women lived on the base. Some women manufactured munitions in the factory, while others tested medium- and long-range artillery in the testing fields.

Women slept two to a room, and Marlyn and Ruth volunteered to room together. Betty and Jean agreed to room together, and Kay was assigned to "a complete stranger" who worked on a different part of the base.[12]

The entire building consisted of thirty rooms, and the women had a large living room for greeting their guests. There was a large resident bathroom with eight sinks, four shower stalls, and several toilets. There was no kitchen, so for their meals, the women could eat with the soldiers or head off base.

The five women settled in, wondering what their classes would be like. The next morning they woke up, showered, got dressed, and found their way to a commissary, one of the Proving Ground cafeteria-style eating areas, for breakfast. It was quickly apparent that there were few women in the dining room, and they appeared to be outnumbered thousands to one (the actual numbers were closer to 6,000 to 1). It was a little overwhelming, and they looked forward to taking the shuttle across the Proving Ground, back to BRL and disappearing into their classroom.

There they found a tall, slender instructor named Sergeant Johnson and a short IBM maintenance engineer, known to all as "Smitty."[13] These two men would be with them for the summer and play key roles in teaching them how to use the IBM card reader, card punch, and tabulator.

Betty already knew a lot about IBM punch cards, but the other women did not. She remembered her days at the *Farm Journal* and the feel of the cardboard cards, 7⅜ inches wide by 3¼ inches high, used to run the journal's survey results. She knew how easy it was for someone to mistype a number onto the card and for the results to be misinterpreted. Now she would learn how to use the cards herself.

The first major component that the two men showed the five women was the plugboard, a twelve- by eighteen-inch panel used to control the IBM card reader and card punch. The plugboard had about 500 holes, known as *connectors*, that were wired to be connected in a certain order. Once inserted into the IBM card reader, the wired plugboard delivered signals to the IBM card reader to communicate the format of the data on the punched cards.[14] For example, how long were the numbers and in what column did they start? For the IBM card punch, the plugboards did the opposite, signaling in what rows and columns the outgoing data would be printed.

Sergeant Johnson and Smitty were good teachers. The women divided up to work on their plugboard exercises. Generally, Ruth and Marlyn were in one group, and Betty, Jean, and Kay in the other. "[W]e learned everything about all the IBM equipment that was there... and we had a marvelous time doing it too," Kay reflected later.[15]

It was all very interesting, but one day in class, Betty was upset. They had a third instructor from time to time, Miss Masincup, and Betty did not understand her explanations. "Finally I got to the point [where] I didn't understand a thing she was saying."[16] This was very frustrating to Betty, who liked to fully understand every technical detail.

Exasperated, she turned to Smitty for help. She asked if she could borrow his IBM maintenance manual, but he could not allow her to take one of the base's only copies. He would be fired if the valuable item got lost. Betty was disappointed. But one Friday he came in and said to Betty, "I'm going away this weekend," and pointed out that he was leaving the technical manual on his desk.[17] He hoped it would still be there, in the very same place, on Monday morning.[18]

After he left, Betty interpreted the statement as tacit permission to borrow the treasured manual. She spent the entire weekend poring over it, studying its explanations and diagrams, until she had answered her questions about how to use the plugboards and wire the circuits for the work she needed the card reader and card punch to do. She barely looked up.

That Monday morning, Smitty arrived in the classroom and found his manual just where he'd left it. He smiled and Betty did too. She felt much more assured about her work. She had her own way of learning technology, and it would not be the last time she taught herself something when she did not understand the way others were teaching it.

Whatever she learned, Betty learned it well because Ruth later recalled that "the only person who ever figured out how to do" advanced functions on the plugboards "was Betty Snyder and nobody else could ever figure out how to do it."[19]

Nonetheless, everyone learned to do the basics—how to use plugboards to set up the card readers and card punch, and how to move cards from the card punch and set up the tabulator for printing them. They would return to Philadelphia knowing a lot more than when they had left.

Some of the men in BRL took the women around to see how they tested artillery to provide input to the trajectories the women calculated at the Moore School. They showed the women the small firing range in the basement where they test-fired small guns and taught them that when they heard a siren in their classroom and a loud bang, they should not be alarmed, that it was just a set of tests in the basement about to go off.

Other men took the five women outside to the firing ranges of the Proving Ground to see how they tested artillery, including the great howitzer guns, and they showed the group how they took measurements to be used for input in the trajectory calculations. "They had sensors at various heights and distances along the path of trajectory," Jean learned. "These sensors were tripped as the shell passed, and the precise height, distance, and time elapsed since the firing was recorded."[20]

One day, Betty and Jean went outside to work on a plugboard exercise and wandered into a nearby field. It was a beautiful summer day, and it was nice to work in the sun. "I remember that we were tracing the wires," Betty recalled, and she became so intent on the work she did not hear the whistle blow to warn that testing was about to begin. Suddenly, and seeming right over their heads, they "starting firing... I thought the place was coming down."[21]

Betty went into hysterics and Jean quickly got them both off the field. Now they knew what a Proving Ground really was.

Surrounded by Vultures

It was not all work on the Proving Ground. Jean and Ruth, in particular, decided to have a little fun. Jean remembered: "As we were working on wiring up plugboards one day, two Marine sergeants and an Army sergeant came into the room and began talking to us... Eventually, the second Marine invited Ruth out, and the Army sergeant asked me out."[1]

Thus began a series of enjoyable dates at the noncommissioned officers' club, "which had booths where we could sit and order food and drinks, and a jukebox so we could dance to records—'A String of Pearls,' 'Don't Sit Under the Apple Tree,' and other Glenn Miller tunes."

Dancing was never Jean's favorite activity, despite her other athletics, but she danced a little and for a while enjoyed relaxing and talking for hours.

But soon the two women found dating difficult, as "the masses of soldiers were intimidating."[2] In fact, with 27,000 soldiers, mostly young enlisted men, almost everywhere the five women went, aside from BRL and the dormitory, they were the subject of stares—and often the objects of catcalls. The fun of being with the young men wore off. "Everywhere we went, we felt like a piece of meat surrounded by vultures," Jean recalled the women thinking.[3]

Instead, the women turned to each other for companionship and friendship. With their dormitory having no cooking facilities, they chose to find a quiet spot off base for dinner. They dropped their notes in their room and took a shuttle close to the gates. They found a favorite diner and took a "little train line" from the base to the town of Aberdeen for a quiet night of long conversations, just the five of them.[4] Over long, leisurely meals, and then, back in the dormitories, they talked about everything: their ideas and opinions about right and wrong, the nation's actions during the war, their family backgrounds, and what they might do on Project X when they returned to Philadelphia.

Betty had thought to bring her blender from home, and she mixed frozen daiquiris that they drank as they talked together in one of their rooms, which they rotated.

"[T]he group was absolutely fascinated with each other," Jean recalled.[5] They engaged in long talks late into the night that Betty called "bull sessions."[6] The five comprised a pretty diverse crowd in terms of family background and religions. Kay, Marlyn, and Ruth came from immigrant families while the families of Betty and Jean had been in the United States for over a century. They comprised a melting pot of religions: Kay was Catholic; Marlyn and Ruth, Jewish; Betty, Quaker; and Jean, Presbyterian. Kay was enthralled with it all. "We discussed everything," she remembered. "We were very complete in ourselves."[7]

On Thursday nights, when the department stores in Baltimore stayed open late for the many women working full-time in the wartime industries, the five often took the train thirty miles south for dinner and shopping. Until Baltimore, Jean had never seen or eaten a lobster. She delighted in wearing a bib, using nutcrackers and tiny forks, and dipping the pieces in cups of butter. "I soon found myself nibbling away on those tiny legs exotic (sic) and loving the taste of it," she remembered.[8] The other women, of course, were amused and enjoyed the experience of Jean's discoveries right alongside her.

It was time to show Jean the wonders of the East Coast. One weekend, Ruth took her to New York City, and when Jean spotted the Empire State Building—102 stories and the tallest building in the world at that time—she could not believe she was seeing it with her own eyes. They had a drink at the famed Brass Rail restaurant, covering an entire city block, and went to Radio City Music Hall to see the Rockettes.[9] Certainly no auditorium in Stanberry could hold 6,000 people. After each day's adventures, they took the subway out to Far Rockaway and had long discussions with Ruth's parents, and Jean felt very welcome.

Next, Marlyn played tour guide and took Jean to Washington, DC. They covered all the key places, including the Capitol, the Jefferson and Lincoln Memorials, and the Washington Monument. They wanted to go to the top of the Washington Monument, but it was closed during the war. They also went to Arlington National Cemetery, and Jean was deeply moved by the rows and rows of white graves, in straight lines. She thought of her brothers, brothers-in-law, cousins, friends in the service, and Joe, and prayed, *Please don't let them end up here.*[10]

Some weekends, Kay and Betty took Jean with them as they headed home. At the McNultys, Jean was astonished to find that as soon as Kay walked in, she slipped into an Irish brogue so thick that Jean could barely understand a word she was saying. The family's welcome to a young woman from halfway across the country was warm, as it was at the Snyder family.

When it came time to leave the Proving Ground at the end of July, the women were wiser and more mature. They had a better understanding of how their trajectory calculations fit into the larger picture of artillery gun testing and use, and they had a better sense of the noise and terror soldiers faced on the battlefield with artillery and other fire around them.

All five were set to return with their mission accomplished: They had learned how to use the IBM card readers, card punches, sorters, and tabulators.

And Jean would return with one more mission accomplished. She was a woman out of debt, having repaid her sister and her aunt Gretchen with money she saved from her salary and some of the per diem money the Army gave them for daily living expenses.[11]

The bonds they formed that summer were strong. The five women had arrived at the Proving Ground as individuals. Now they were leaving as a team.

The Dean's Antechamber

In late July 1945, the women returned from Aberdeen triumphant in their new knowledge of the IBM machines. They were excited to move on to Project X, but they were met with silence. No one gave them their next assignment, not Herman, not John Holberton, no one.

"They really didn't know what to do with us," Jean recalled.[1] Being resourceful, they found the only space available to sit in a Moore School still filled to overflowing with Army projects. They took over the dean's own waiting room, setting up desks in what was known in those days as an *antechamber.*

It was a little baffling, as Kay and Betty had been working six days a week for the Army computing project since 1942, and Marlyn and Ruth since 1943, and now when they knew a major project awaited them, there was no guide or introduction. At least not yet.

But it was not their job to speed up an Army assignment, and they knew they would hear when someone needed them. Their job now was to wait. They had no way to know that the men were working hard to finish ENIAC, already months behind schedule, due in part to difficulties obtaining necessary supplies.

People came in and out of the antechamber to visit the dean, and a few days after the women arrived, two men they did not know came. The women said hello, and the men stopped to respond. "They were introduced to us as Dr. Stan Frankel and Dr. Nick Metropolis," Kay said, "but otherwise nothing else was said to us at all."[2] With so much military secrecy around them, the women knew not to ask questions.

As the women waited, the country waited too. By the end of July and the first week of August, the entire country was holding its breath wondering when the main invasion of Japan would begin. After years of battles with horrible losses for both sides, island by island, from Hawaii across the Pacific, the US forces sat in Navy ships at the doorstep of the Land of the Rising Sun.

On July 31 and August 1, the *Philadelphia Inquirer* reported that senior US military officers had warned twelve Japanese cities along the cost to evacuate their citizens before US bombers arrived[3] and set fire to a long stretch of coast only eighty miles from Tokyo.[4] Everyone knew the end of the war was near, but no one knew exactly when.

The five women read the newspapers daily and listened to the radio news broadcasts closely, along with nearly everyone in the country. By Sunday, August 5, the US military warnings became "surrender or die" to Japanese cities extending "the entire length of the Japanese homeland."[5]

Although a large number of US forces massed for the invasion of Japan,[6] no one thought it would be easy and Japan showed no signs of surrender. On August 5, the American public received Japan's strong official response:

> Despite the tremendous weight of warplane and warship bombs and shells being heaped on the homeland, Tokio (sic) declared that it would never be forced to its knees by air power and was ready to meet any Allied invasion.[7]

Reports obtained by the US military showed that Japan was training its civilian population—men, women, and children—to defend themselves, all while building defenses across the country that Japanese officials declared were impenetrable. It was clear to the public that an invasion of Japan would be very costly in terms of lives, for the United States and Japan. But the public did not know that the internal US military projections showed a likelihood of a million US casualties and millions of Japanese civilians.[8]

As Kay sat in the antechamber, she was worried for her brother Pat. Jean and Betty were worried for their brothers. Marlyn and Fran worried with them for other family, friends, and all US soldiers. The entire country waited.

Then, on the morning of August 6, the women and the country awoke to startling news. That day, in the center of the city of Hiroshima, Japan, the Army had dropped a secret weapon, an atomic bomb. The *Philadelphia Inquirer* proclaimed the earth-changing event with the headline: ATOMIC BOMB, WORLD'S MOST DEADLY, BLASTS JAPAN; NEW ERA IN WARFARE IS OPENED BY U.S. SECRET WEAPON.[9] A one-square-mile section of downtown Hiroshima, a city of 350,000 people, had been destroyed.

Many expected Japan would surrender immediately, but that did not happen. On August 9, the newspapers and radio broadcasts announced another horrific bombing: "The world's second atomic bomb, most destructive explosive invented by man, was dropped on strategically important Nagasaki on western Kyushu Island at noon today," the front page of the *Philadelphia Inquirer* wrote.[10] Japan surrendered.

The five women were relieved and horrified. On the one hand, the war was finally over and the soldiers would be coming home. Millions of lives had been saved on both sides. On the other hand, they did not understand why the targets of the terrifying weapons had to be civilian populations. Kay thought it was a "terrible, terrible way of ending":

> [I]f you're going to use the bomb, why not use it on all these big naval installations which Japan had, because they were a Navy power? . . . [W]hy didn't we destroy naval installations?[11]

Jean was shocked by the devastation caused by the bombs: 80,000 to 200,000 casualties in Hiroshima alone. "The reality of the atomic bomb was an unimaginable horror," she shared later, "and, although our country was the one that had dropped them, the Philadelphians I knew were afraid of its awesome and horrific power."[12] She was upset when she learned that President Harry S. Truman said he'd slept well the night he made the decision.

Betty was distraught. As a Quaker, she had a hard time facing the scale and scope of the destruction, particularly for women and children. Unlike President Truman, she could not sleep for days.

Japan agreed to an unconditional surrender with one provision: Emperor Hirohito would remain on the throne. On August 14, Emperor Hirohito broadcast a radio address to his country for the very first time. He told his people that "the enemy" had used "a new and most cruel bomb, the power of which to do damage is, indeed, incalculable, taking the toll of many innocent lives." He predicted what would happen if Japan did not surrender. "Should we continue to fight, it not only would result in an ultimate collapse and obliteration of the Japanese nation, but also it would lead to the total extinction of human civilization."[13]

It was the first time the people of Japan had heard the Emperor's voice. They, like he, would unconditionally surrender to the Allies and they, like he, would follow the orders issued by General Douglas MacArthur, the newly appointed Supreme Commander for the Allied Powers. The long war was over at last.

On August 15, the Philadelphia newspapers had a one-word

headline: PEACE.[14] In Philadelphia and across the country, celebrations erupted in the streets for Victory over Japan Day, or V-J Day. The five women in the antechamber rushed out to celebrate with other Computers and hundreds of thousands of people in Philadelphia.

John Holberton, Kay, and Alma, Kay's sister-in-law and Pat's wife, celebrated together downtown. "There was wild cheering, singing, dancing and laughing," Kay said. "Everyone was grabbing everyone else and kissing them. We were all so happy that the war was finally over. We were out of our minds with joy."[15]

Crowds gathered everywhere, women and men dancing in the streets, men playing trumpets. There were so many people that trolleys and cars could not pass. Four and a half years after Pearl Harbor, the soldiers could come home and life could go back to normal. While many women were concerned about losing jobs they enjoyed, these five women felt their positions were secure, even if they were not quite sure what their next job would be.

After V-J Day, the mood at the Moore School changed quickly. Other Computers, having completed their wartime work, packed their things and made arrangements to go home or move on to new job opportunities in Philadelphia and elsewhere. Ruth said good-bye to Gloria Gordon, who had been her roommate and was returning home to New York City to work in the Brooklyn Navy Yard. Marie Beirstein, formerly on Betty's computing team and then on a differential analyzer team, accepted a transfer down to Aberdeen Proving Ground to work on BRL's analyzer.[16]

Soon the only women left were those connected with the ENIAC, and it was getting very quiet in the Moore School.[17] But the five women had not been dismissed or reassigned. Unlike the rest of the country, now moving on to new postwar activities, the five women were still waiting.

Finally, one day in August, Arthur Burks popped into the antechamber. Kay knew him because he married Alice Rowe on her team in the analyzer room, but no one else knew him. Arthur was a PhD logician with a real knack for electronics. He had come to help the Army's war effort and wound up on the ENIAC team.

His hands were full of large, white papers, rolled up in long tubes. Arthur wanted to talk with them, but they needed a space to unroll the big diagrams.

The five women looked at each other with excitement. Could this at last be the next assignment they had been waiting for?

A New Project

Kay, Marlyn, Ruth, Betty, Jean, and Arthur found an empty classroom with a big wooden desk on the second floor and put down "blueprints, the wiring diagrams, and block diagrams" of a very large computer he called ENIAC.

Arthur explained that Project X was one of the names for the project to build ENIAC. He pronounced ENIAC with a short *e*, and said it stood for "Electronic Numerical Integrator and Computer." It was the computer being built in the lab with the RESTRICTED sign at the back of the first floor of the Moore School. ENIAC's engineering team was led by Dr. John Mauchly and J. Presper Eckert, and Arthur was on it, along with a number of young engineers the women had seen in the hallways.

He explained that in addition to the ENIAC hardware, the contract between BRL and the Moore School called for ENIAC to be delivered with a working ballistics trajectory program. That was now the women's job, and further, ENIAC would have to run the trajectory calculations at blindingly fast speeds to fulfill the promise that John and Pres had made to BRL.

After all, BRL had spent enormous amounts of time, money, and resources not to build a modern computer but to dramatically reduce the time to create firing tables. Just because the war was over did not mean BRL was rolling up its carpets and going home. Veblen had kept the trajectory computations going between the First and Second World Wars, and while no one wanted another war, BRL planned to continue its mission. If the Army created a new howitzer, BRL would give it a firing table. ENIAC, if it worked, would play a key role in BRL's future.

The five women paid close attention and understood their new project: Preparing a ballistics trajectory program for ENIAC, the "acceptance test" for this computer still in test mode, would be their assignment. It would become "one of our missions," Jean recalled.[1]

Although the war was over, and the project was rather astounding, none of the women considered saying no. First, they were still in wartime mode. If the Army needed them, they would help. Second, the assignment made sense to them. They were, after all, the subject matter experts on BRL's trajectory equations. BRL had invested a lot of time in training them on graduate-level numerical analysis and trajectory-calculating methods. Collectively, they now had years of experience. It made sense that Herman and John Holberton would add them to the ENIAC team.

That they were being asked to do something no one had ever done before—learn to program the world's first general-purpose, programmable, all-electronic computer—did not give them pause. As Kay later commented, "Lots of people were doing out of the ordinary things."[2]

What the women did not know at the time was their efforts, as Computers, had been very successful. During WWII, American artillery became known for being both deadly and incredibly accurate. Ernie Pyle, a well-known WWII war correspondent, had written in 1944, "Our artillery . . . The Germans feared it almost more than anything we had."[3] BRL aimed to keep things this way by continuing the calculation of trajectory equations and the production of accurate firing tables, hopefully with ENIAC.

But first they had to learn how ENIAC worked, and that meant looking more closely at Arthur's diagrams. And Arthur had lots of engineering diagrams to share with the five women. Of course, it would have been easier if he had brought a programming manual, but although Adele was writing one in the background, it was not available to the women and would not be published until the following year, June 1946.[4] The women were a little surprised that "they had no books or anything to teach us," Jean remembered.[5]

First Arthur unfurled a circuit diagram and explained it was a drawing for engineers. It showed "how this tube activated this tube which activated that tube,"[6] Kay learned. He explained that the vacuum tubes were the main electronics component of ENIAC. There were 18,000 of them in ENIAC, and he confessed that they often burned out.

The circuit diagrams introduced material that was daunting, yet exciting. Betty learned to read a blueprint diagram from the lower left-hand side, "where the circuit takes the energy," across the page to the right following the energy as it passes through gates, some of which allow the energy to flow through and some that stop it.[7]

Then Arthur opened a second diagram that he called a block diagram and walked them through it. Block diagrams, they learned, were different than circuit diagrams. They show how the functions of a unit or system interrelate. Betty learned that a "block diagram is something that is not used by the engineers or the people who wire the machine together," but "mostly after [the computer] is all built . . . [it's] one step higher in learning."[8]

Finally, Arthur next unrolled what he called a *logical diagram*. It showed the front of an ENIAC unit. The five women leaned in. This diagram indicated places for dials, switches, plugs, and cables on the front of the ENIAC units. These switches and plugs, Arthur explained briefly, would be how they set a unit to do what they wanted it to do and to communicate with the other units of ENIAC. These diagrams were the key to "the logistics of the whole machine," Kay discovered.[9]

Kay appreciated the three types of diagrams and thought they provided a good overview. "We learned... you might say, from the back forward. We learned all about the tubes first, and then came around and found out what the front did."[10]

But all too quickly Arthur had to return to the ENIAC room. In his introduction, he had spared all the time he could. Now he needed to go back and continue testing ENIAC. The women would have to work together to teach themselves how the forty units of ENIAC worked and how to program them.

In passing, Arthur noted that unfortunately, the women did not yet have the security clearance to enter the ENIAC room (which was different from their clearances as Computers, and Kay's additional clearance to enter the analyzer room). If they saw him in the hallway, they should feel free to ask a question or two, but they could not venture into the ENIAC room to see the computer or ask the engineers about it.

Instead, Arthur indicated the diagrams once more.[11] Study them, he urged as he left a million questions unasked and the five women shaking their heads.

This situation put the women in a bind. "We didn't know that there would be so little documentation that would be available to us."[12] Kay felt a bit lost.

Jean felt the same way. "So they gave us these great big block diagrams... and we were supposed to study them and figure out how to program it and how the thing worked. Well, obviously, we had no idea what we were doing."[13]

But Marlyn had a sense they would grope it out and figure it out together.[14] She turned out to be right.

Divide and Conquer

Alone in the classroom, the women pored over their material. Such odd names: accumulator, high-speed multiplier, divider and square rooter, initiating unit, cycling unit, and more. They decided to divide up the units, pairing off on certain things and coming back to teach each other. A divide-and-conquer approach.

Which units did they want to start with? Jean and Betty offered to take the accumulator. Kay took the high-speed multiplier unit.[1] Marlyn and Ruth probably took the divider and square rooter, a third unit dedicated to mathematical operations.

The pairs dug out their units' diagrams from the stack, said goodbye, and went to find some open space. Now that the war was over and there weren't Army projects in every nook and cranny of the Moore School, there was much more space. Betty and Jean found an open classroom on the second floor, albeit a little noisy due to some construction above them.[2] Marlyn and Ruth staked out the old fraternity house at 32nd and Walnut, now quiet after the departure of Lila Todd's computing team. Kay stayed in the antechamber, continuing to greet guests and do her work on one of the tables.[3]

Kay was happy, and surprised, when Fran showed up. Fran had continued working on the analyzer when Kay went off to the Proving Ground, as Herman and John Holberton did not want to lose both of their supervisors. But they wanted Fran on the ENIAC team. After she helped wrap up work on the analyzer, sending results and confidential trajectory documents to BRL and transitioning the analog machine back to the Moore School, they asked if she wanted to continue working for the Army on a new project. When Fran said yes, Herman and John Holberton moved her to the ENIAC team.

Kay quickly filled Fran in on the summer in Aberdeen, and the basic information she had learned from Arthur about reading the ENIAC diagrams. The two put their heads together to study the high-speed multiplier and figure out how it worked. The two best friends were back together.

They stared at the multiplier's main panel, with five rows of eight switches neatly arranged in the middle of each of the three multiplier units and small round connector plugs beneath, and tried to figure out how everything worked.

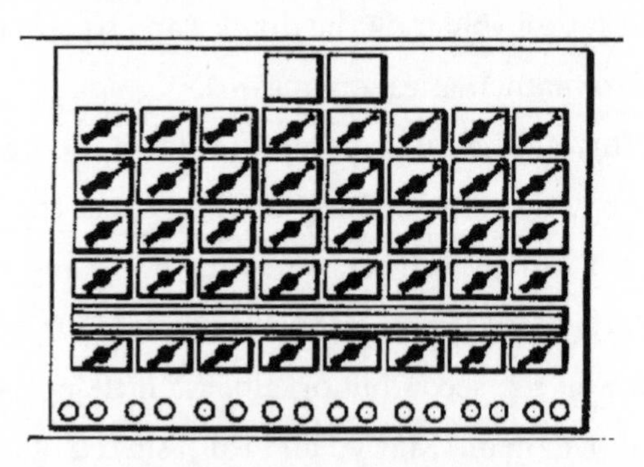

Forty switches in middle of front panel of high-speed multiplier
ENIAC Patent 3,120,606, Sheet 37[4]

Meanwhile, Marlyn and Ruth were trying to figure things out at the fraternity house, and the answers did not come easily: "It was frustrating to us to not be able to get the answers the way we wanted them. However, I think we were aware that this was something new and that [it] was an experiment. We had to be patient."[5]

Jean and Betty sat in a classroom on the second floor and began to study the diagrams of the accumulator.[6] They glanced a bit at the blueprint and wiring diagrams on the back of the accumulator, but what interested them were the diagrams of the front. The block diagram and logical diagrams indicated a long rectangular unit with four main sections there.

At the very top was a ten-by-ten grid. Below that, there seemed to be places to plug wires with large, wide rectangular heads. The middle part of the front panel had twenty switches in three rows: four switches on top and two rows of eight switches each below.[7] Twenty small round plugs beneath the switches seemed available for some type of special connection.

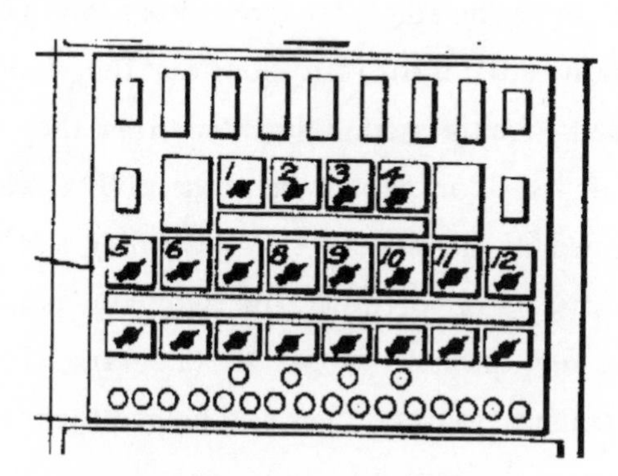

Switches in middle of front panel of accumulator
ENIAC Patent 3,120,606, Sheet 25

They puzzled over the diagrams for some time. Meanwhile, the noise in the background was almost deafening, as the Moore School was adding on a third floor and the jackhammers were going full blast.[8] Although the August days were hot and humid, the two women could not decide if it was better or worse to open the windows. Either way, it was very noisy.

They sat staring at the large rolled-out sheets and trying to figure them out. They sat with their two college chairs, "the kind with an armrest that curls around into a small writing desk," with the diagrams spread across the two desks and their heads close together trying to figure everything out.[9]

One day, with jackhammers pounding away and dust falling on the women's heads, a tall, slender, bespectacled man came in and walked around, looking up at the ceiling intently. It took him a moment before he realized that anyone else was in the room, and when he did, he smiled and introduced himself. He was John Mauchly, and he was "just checking to see if the ceiling was falling down."[10] Neither Jean nor Betty had met him before, but they knew who he was and they were excited to meet him, as he and Pres were almost "mythical figures" to them.[11] He explained that if the third floor that the Moore School was building collapsed, then the second-floor classroom they were in might collapse onto the first floor below, where the ENIAC was being built. He seemed satisfied for the moment that the ceiling was holding.

Not one to miss an opportunity, the two women asked him if he would answer a few questions about the diagrams they were looking at. A born teacher, John said yes and helped them to learn some advanced features of the accumulator.

The accumulators, and there were twenty of them, could do more than just add and subtract numbers. "Each of the twenty accumulators could receive and store a ten-decimal-digit number,"[12] Jean learned, and a sign for the number, positive and negative. This seemed like a very useful feature to Betty, who had spent so much time during the war writing down the result of every calculation on the desktop calculator and retyping it back onto the desktop calculator for the next calculation. Now interim results could be stored in the accumulators until they were needed later in the program—temporarily.

Like Arthur, John could stay only for a moment, as he had classes to teach and ENIAC testing to finish supervising. He told them, however, that he shared an office with John Holberton next door. If they needed him and he was around, he would be happy to answer their questions.

About two weeks later, the six women reconvened to share what they had learned from their diagrams. They were excited to regroup. First, Kay introduced everyone to Fran and the other four women welcomed her warmly. By Herman's assignment, Fran was now a permanent member of their programming team. Then the six women rolled up their sleeves and began to teach each other the units of ENIAC.

In this first pass, they figured out the "arithmetic units," or those dedicated to mathematical functions. Betty and Jean taught the accumulator and how it could be used for addition, subtraction, and temporary storage; Kay and Fran taught the high-speed multiplier and how it could multiply large numbers.

Ruth and Marlyn likely discussed the divider and square rooter, the third arithmetic unit of ENIAC. If so, then they would have shared that this unit had four rows of eight switches in the middle of the front panel, and needed to be set to receive the numerator and denominator for division or a radicand, the number from which a square root would be taken. Jean and Betty shared their further discovery that accumulators provided a valuable function, storage of temporary results. But the storage was very limited. There were only twenty accumulators, so the maximum amount of temporary storage at any given time was only twenty numbers.

Not a bad start to their collective work. It was time to celebrate their new knowledge and process. Lido's Restaurant, on Woodland Avenue, was just the place. "[W]e squeezed into a booth...[at this] Italian restaurant that was somewhat dark and always crowded [with a] festive atmosphere."[13] It was fun to be back together again.

The five women tried to include Fran in their discussion and draw her out, but she stayed close to Kay and quiet. Marlyn wanted to know Fran better, but knowing how difficult shyness could be, she respected the distance Fran wanted to keep:

> Fran was very quiet...She kept to herself more than the others did. She did what she was supposed to do...[S]he didn't mingle that much with everybody...[S]he was very private, very quiet.[14]

The next day they resumed work. There were more units to learn of ENIAC—units that controlled the storage of numbers, the powering on and off of the unit, and the pulse that circulated the computer when it was running a program.

For their next unit to learn and then teach, Betty and Jean chose the function tables. These unusual units had two square faces full of dials: Each face had twenty-six rows of twenty-eight switches, for 728 switches per side and 1,456 switches per unit.

To learn more about these units, Betty and Jean buttonholed Bob Shaw in the hallway. They heard he was the engineer who designed and built the function tables, so he could help them understand it.[15] While the two women could not go into the ENIAC room, there was no rule against approaching anyone outside of it. Bob was a warm individual, prone to practical jokes and happy to help them.

Jean and Betty learned the special role of the function tables in

ENIAC's ballistics trajectory calculations. The trajectory equations had many "constants," numbers fixed for the calculation, and the function tables were designed to hold these numbers. Jean and Betty could set the switches on the function table for each digit of a number, turning each switch from 0 to 9.

Today, we would call the function tables "read-only memory," because the numbers, once set in the switches, could not be changed during the course of the program. Now Jean and Betty felt they were making progress! They could not wait to get together with their team to share and teach.

Over time, the team learned about four types of units:

1. Arithmetic units that they knew: accumulator, high-speed multiplier, and divider and square rooter, for addition, subtraction, multiplication, division, and square roots.
2. Function tables for read-only memory and the accumulators for temporary memory.
3. "Governing units" that controlled other units and, in particular, the "initiating unit" that turned power on and off to the rest of the units, and the "cycling unit" that issued a program pulse to start the operations of other units.
4. Input and output units of ENIAC worked with the IBM equipment the women knew so well. ENIAC's "constant transmitter" unit was paired with the IBM card reader to read data from punched cards and send it to other units for calculations. ENIAC's "printer" unit was paired with the IBM card punch and sent the results of ENIAC calculations to be punched onto cards. These cards could be taken to the IBM tabulator and printed onto accordion-style paper for people to read.[16]

The women found the "cycling unit" especially interesting. It issued a pulse that served as the "GO!" switch for other units of ENIAC and started the activity of the other units. The cycling unit issued its pulse in a regular pattern, making it the heartbeat of the great ENIAC.[17]

By late September, life in Philadelphia was returning to normal. Stores sold fresh produce, and fragrant Pennsylvania fall apples filled the shelves. Garages again offered abundant gasoline and stocked new tires for cars. The six women reassembled again and again to share and teach what they had learned. With each pass, they grew in confidence. As they learned new units by themselves and from each other, they grew as a team.

As men returned from war to resume their jobs in the factories and fields, in the fall of 1945, the government ran campaigns encouraging women to leave their positions to make room for these soldiers. Some women were happy to be at home again; others returned only reluctantly because they had enjoyed their work and their income. But the six women of ENIAC, the "ENIAC 6," were different. No returning soldier had their skills and no returning GI could replace them.

The two "master programmer" units still puzzled them. Each had a dizzying array of switches on the main panel in the middle of each unit. Four switches spread across the top row, then six rows of ten switches, and five switches across the bottom row. Beneath the bottom row of switches, fifty-five small round circles seemed ready for special connections. But these special units would have to be left for another day.

They knew what most of the units of ENIAC did, and it was time to move on to their next task. Each unit of ENIAC did one or a few things well, but how could they make all the units work together?

A Sequencing of the Problem

The six women needed to figure out how to communicate a human problem to the complicated computer. By now, with wartime projects wrapped up, there was more space in the Moore School and John Holberton gave them an office of their own on the second floor. Having a single space to work together would make this next task easier. They sat down to figure out how to program a problem on ENIAC.

Today people learn programming from books and classes, online courses, and activities. Modern programmers also use tools such as programming languages, operating systems, and compilers. None of these courses or tools existed in fall 1945.[1]

But no one taught them at the Moore School. The six women had to find clues in the drawings Arthur had left. "How we learned to program," Betty said, "I have no idea... [W]e must have thought it up ourselves."[2]

It took a few steps of induction, Kay remembered, and when thinking about the diagrams of the front of the units, "we had to figure out how you would put a problem on, how you get the machine to do what you wanted it to do. And you had to do that from the front by setting switches and plugging in plugs and things of that sort."[3]

The six women realized that they would have to string the units of ENIAC together to pass numbers from one unit to another for multiplication, division, and the other operations. On ENIAC, numbers were called *digits*, and these ten-digit numbers passed between units through "digit wires" and along "digit trays."

They also needed to move the program pulse from the cycling unit—the GO! of the ENIAC—to each unit to start its operation. The cycling unit pulse ran along the back of the units and could be used by Programmers to start one unit or multiple units for the next step or steps of the program. When a unit finished its operation, it issued a program pulse of its own that could be sent to another unit to start the next operation.

Betty summarized their discovery as learning that programming of ENIAC meant "[w]e had to hook the machine together... to give you sequencing." It would be entirely up to them to handle each and every logical and physical step of their program, and to prepare the entire "sequencing of a problem."[4]

Tom Petzinger of the *Wall Street Journal* would later call the ENIAC 6 the "operating system" of ENIAC:

> Running the ENIAC required setting dozens of dials and plugging a ganglia of heavy black cables into the face of the machine, a different configuration for every problem... Every datum and instruction had to reach the correction location in time for the operation that depended on it, to within 1/5,000th of a second.[5]

Writers working for BRL a few years later would use similar terms. One called this method of "programming by means of pulses, switches, and cables"[6] and another labeled it "direct programming."[7] The original direct programming method fascinates computer scientists to this day.[8]

There would be many more details of the ENIAC to be worked out, but the women were well on their way to learning the basics and nuances of how the units worked, and the system of connecting them together to solve a problem. What was clear was that programming the ENIAC to do their ballistics trajectory equation would be "no mean feat."[9]

As they studied the drawings closely to learn about the interactions of the units with each other, Kay had a eureka moment. She realized that a key feature of the master programmer units was to run "loops." Its two units had switches that could be set to allow calculations to be run multiple times; for example, a switch could be set to allow a calculation to run five times across the same sequencing of wires and switches. Each calculation would use the results of the prior one.

Today programmers call this process a *loop*, and it involves reusing code. On ENIAC, the master programmer loop allowed the ENIAC 6 to reuse previously set wires and switches. "Kay was very creative," Jean remembered, and she "was the first one that made me see how powerful" the loop technique was.[10] It was a milestone for the team.

All this time, they still had never seen ENIAC and they wondered when they would be allowed into the room.

Having learned the units of ENIAC and its "direct programming" method of programming, the ENIAC 6 began to think about their ballistics trajectory program. How could they break it down? What steps would be required?

According to Kay, this process quickly came to a halt:

> We had barely begun to think that we had enough knowledge of the machine to program a trajectory when we were told that two people were coming from Los Alamos to put a problem on the machine.[11]

That was the day in mid-November when Herman suddenly appeared in their doorway. He had a serious look on his face. It was clear he was about to issue a command. "Follow me," he announced. "We need you in the ENIAC room."

The women would return at some point to resume their ballistics trajectory work. But for the moment, he had an even more pressing problem for them.

The women looked at each other with excitement. This was the moment they had been waiting for. Their exile from the ENIAC room was over. With big smiles, they jumped up and followed Herman.

A Tremendously Big Thing

For the first time since Kay and her analyzer team saw the two-accumulator test, the women saw ENIAC. It had changed completely from what Kay saw a year and a half earlier. ENIAC was now eight feet tall, eighty feet long, and arranged in a huge U to fit into the large room. Sixteen towering units on the left and sixteen towering units on the right, with eight towering units before them in the middle. The units were pushed out from the walls so that people could access them from the back and the front.

There were other people in the room, but initially the women only had eyes for ENIAC. It completely commanded their attention.

Wow, thought Marlyn. *We're never going to make this do anything. It's too big.*[1]

Jean was thrilled. It was more magnificent and intimidating than she had imagined.[2]

Betty found ENIAC a bit ominous: "[It] was a tremendously big thing...[I]t took a whole big room and [it was] black and dark as could be."[3]

Ruth marveled at the number of tubes (18,000) and the imposing size of the room.[4]

It was perhaps Kay who best summed up the women's feelings: "We had never seen it, all assembled before, and there were all forty big tall black units just standing around looking at us . . . [and] we were delighted."[5]

They walked around the units, examining each one individually: the high-speed multiplier, the divider and square rooter, the many accumulators. The diagrams they had studied so closely came to life before their eyes. It was a moment they would remember for the rest of their lives.

Herman gave them a minute and then pulled them out of their reverie.

He cleared his throat, and they remembered that there were many other people in the room: Herman and Adele, John Mauchly and Pres Eckert, Arthur Burks, Bob Shaw and the young ENIAC engineers they had seen in the hallways.

The two men from New Mexico were also present; they had met briefly in the antechamber in late July. Now Herman could introduce them properly. They were Drs. Nicholas Metropolis and Stanley Frankel of the Army's Los Alamos Scientific Laboratories in New Mexico, an installation that had recently been revealed to the public as the place where the atomic bomb had been created. The women nodded and took a deep breath; they now knew who these men were.

Nick and Stan told the six women that they had brought a calculation from Los Alamos for ENIAC, but it was highly classified and they could not discuss it.[6] Later the women would learn that John and Pres objected to the timing of this use of ENIAC. For two years they had received nonstop pressure from officers at BRL to finish their contract and deliver ENIAC as a working computer with a working ballistics trajectory program. It was the trajectory program that BRL wanted and planned to use when ENIAC was delivered to its permanent home at the Proving Ground after completion. John and Pres feared that this Los Alamos interruption would further delay their contract deliverables. While BRL officers were sympathetic, John and Pres were overruled. The Los Alamos need, whatever it was, came first.

"We're sorry we can't explain more about the problem to you," Ruth remembers Nick and Stan saying to the six women, "but we'll show you how to run the cards and how to run the machine."[7]

The two scientists slipped small, prepared cards into the holders on the front of many ENIAC units. Each card showed the switch and wire settings for the unit. It soon became clear that the women were there as much-needed extra hands to set the switches and string the wires together to weave Nick and Stan's problem onto ENIAC.

The men and women spread across the room and Herman acted as the great conductor. Working off the cards, he directed the wiring across ENIAC. With two people on each end of the long, thick digit cables and the thin, black program pulse wires, he barked the directions.

"Ready! Accumulator One!" Herman cried. "Program Line Input A-0 to Switch Five, set to receive from Alpha!" Then he called out, "Accumulator 2, Program Line Input A-0 to Switch 5, set to add to Alpha, output program pulse to line A-1," and the women and men he directed jumped to set up accumulators 1 and 2.[8]

With each command, members of the group sprang into action, stringing digit cables and program-pulse wires and then setting thousands of switches. The women lifted the long, black, heavy digit cables along with members of the hardware team. They wired the thinner program pulse cables. They carefully set dozens of switches on each unit according to the settings on the card, and they set hundreds of switches on the function tables.

They worked at the same time, in coordinated fashion. No wonder Herman needed so many extra hands.

The women were thrilled to finally have their hands on the computer. "To see the switches!" Jean said. "To turn the switches!" For her, this was heaven: "a physical experience as well as an intelligent, logical experience."[9]

But it was also a little cartoonish. "I mean this is a scene like Looney Tunes," Jean remembered later with a huge belly laugh, when "Herman said 'Rrready,' and "everybody's going to set their switches."[10]

It was a serious moment too. Everyone had a sense of the history they were witness to in that moment. When the ENIAC began to run, and the lights of the accumulators flashed, the women felt a little emotional. They could not say for sure what kind of computers would come after ENIAC, how large they would be, or how people would program them. But they understood this was the beginning of something new, something that would change the world.

The six women stayed in the ENIAC room working with the Los Alamos scientists, John, Pres, and the ENIAC hardware team for a few weeks. Every day was exciting.

At the time, they were told nothing more about the problem they were working on. Years later, they learned they had helped the ENIAC run rough calculations of triggers for the hydrogen bomb.[11] Secretly, and over the objections of some of the atomic-bomb fathers, including Robert Oppenheimer, Los Alamos started to develop the hydrogen bomb—1,000 times more powerful than the atomic bomb. But Los Alamos scientists and mathematicians had trouble designing the trigger for the weapon and needed something to help them find the answer. Dr. John von Neumann, a world-famous physicist and mathematician, was a consultant and visitor to Los Alamos and an adviser to the BRL. He knew about both projects, and hearing the need of the Los Alamos scientists, he asked BRL for use of ENIAC.

Although John and Pres had objected to the timing, clearly this assignment from Los Alamos was very important, and John von Neumann had prevailed on Leslie Simon and Paul Gillon at BRL to allow it. Thus, the two co-inventors had no choice. ENIAC would first be a guinea pig for Los Alamos.[12]

Now with the Los Alamos scientists, the restricted ENIAC room had further secrets. Because no one could see their data, Stan and Nick brought already-punched cards from Los Alamos—about a million of them.[13] Then the six women helped them run the punched cards on their classified "test" program. They were surprised by the high level of responsibility they were given.

Kay and Fran generally worked on the day shifts with Nick, and Betty and Jean generally worked nights with Stan. Marlyn and Ruth helped on both shifts as needed. Throughout the process, Pres and John were constant presences in the room, "like mother hens" to the ENIAC still in testing mode. Kay remembers, "They stayed with that computer night and day, and whatever went wrong, they went in there" to fix it.[14]

Pres tended to work days and John generally took night shift, but the work never fell into regular shifts and days regularly rolled into nights. "Many a night Eckert and I ran the full six blocks to the 30th Street station," Kay recalled, "and then collapsed into a sound sleep on the last train out to the suburbs."[15]

The program generated punched cards that the women then ran through the tabulators to print them out. Nick and Stan made sure to keep the accordion-paper printouts with the results of their calculations securely locked up, first in their locked briefcase and then in a safe.[16] One day, Nick left the briefcase at a nearby drugstore when he and Pres stopped there. When they realized their mistake, the two ran back at breakneck speed and breathed a sigh of relief when the clerk handed the briefcase back to them. "If it had had something of value in it," he quipped, "you guys would have lost it."[17] No secret was more closely held at the time and many of the documents remain top secret today. The ENIAC team, women and men, would not know for years the nature of the equations they had calculated.

After a few weeks of dedication to the Los Alamos project, it was decided that Kay and Fran would stay with Nick and Stan to continue the Los Alamos work. They got to know the two men pretty well, and Kay would keep in touch with them for years to come.

Programs and Pedaling Sheets

Jean and Betty spun off first and went back to the team's space on the second floor. Flush with the new confidence and education of working hands-on with ENIAC, they were ready to dive deeper into the challenge of programming the trajectory. Like everything else, they split up the task to make it more manageable. The initial division of labor was simple: "Jean did the mathematics, and I did the logic," Betty proudly shared later.[1] Jean, the math major, broke down the complex trajectory equations into smaller segments. Betty, the logistician, took these smaller mathematical segments and broke them down further into the small and incremental steps that ENIAC could handle.

As Computers, they relied on their intuition, knowledge, and expertise, but ENIAC had none of those attributes. They realized that what ENIAC needed to do and what they had done as Computers was very different. For a Computer to add 987,643 to 495,145, she would know to type the digits of 987643 onto a Monroe desktop calculator, and then:

1. Hit the addition button (+).
2. Type the digits 495145 onto the calculator.
3. Push down the addition button again.

4. Watch the small wheels turn and see the sum come up in the small windows of the lower level of the bar at the top of the calculator before writing down the sum on a long white trajectory sheet.

But ENIAC knew nothing. It would do only what its Programmers, through careful planning and preparation, made it do. For ENIAC to add the same two numbers required many steps, including:

1. Knowing that accumulator 4 held the result of prior calculations needed for the next step (e.g., 987,643).
2. Setting the operation switch of accumulator 4 to Greek lowercase letter "alpha" to receive a number on the "alpha" digit input connector located above the programming switches.
3. Setting the operation switch of accumulator 6 to "A" to transmit and add its number (e.g., 495,145) to accumulator 4 via its "A" digit output connector.
4. Initiating accumulator 6 to transmit and accumulator 4 to receive on the same program pulse.
5. Keeping the result in accumulator 4 until it was needed later in the program.

In thinking about the scope and breadth of information needed for the large ballistics trajectory program, Betty and Jean realized they needed a notation system that would capture all of the mathematical and logical steps, and all of the physical detail of the trajectory program. They decided to create it themselves, and they knew just where to start—with a big, blank sheet of paper.

As they sat in front of the familiar white sheets, they began to write. Across the top, they drew twenty-seven columns. They labeled the column on the far left *M.P.* for "master programmer"; *Acc 1* and *Acc 2* for accumulators 1 and 2; *Divider* for the divider and square rooter; *Acc 3* through *Acc 10*; *HSM* for "high-speed multiplier"; and so on for most of the units of ENIAC.[2]

Down the left side of the sheet, they drew sixteen rows, one for each incremental step of the program. They knew they would be creating many more pages for the many steps of the trajectory program to come. At the intersection of the rows and columns lay 450 neat, small squares waiting to be filled with every detail of the program's step: its digit wires, program pulse cables, switch settings, and more.

To add more information, Betty used color. She had a four-colored pencil, a Norma pencil, and it helped her to label the sequencing of pulses, switches, and cables through the computer.[3] It helped to show the flow of data across the computer.[4]

The key to working with sheets was to work their way down, one step at a time. To Betty it was a lot like riding a bicycle, pushing one pedal at a time. She promptly named them "pedaling sheets," and Jean enthusiastically agreed.[5]

The two women set to work filling in their pedaling sheets with the mathematical, logical, and physical details of the ballistics trajectory program. One row "for every add cycle of the machine." Betty proudly explained that it "was our idea of a flow chart," and she and Jean invented it together.[6]

During those brain-busting days, Betty and Jean practically lived at each other's homes. Jean spent a lot of time with Betty's family, the Snyders, as Betty was still living at home with her parents. "They lived out in Narberth, and I used to go out there two or three nights a week and stay with them." She played badminton with her younger brother; talked with Betty's father, who she looked up to as a "schoolteacher and astronomer"; and was basically adopted by Betty's mother: "she was a wonderful cook; she was very vivacious, and a wonderful mother."[7] Jean always had a wonderful time.

When Jean's roommate was out of town, the two did things in reverse. "Betty used to come in to Philadelphia and stay with me," and then they explored the city and went out on the town. From the time they started with ENIAC, Jean felt that she "was a team with Betty," who she called "my first perfect partner."[8]

Betty wondered about the master programmer unit. Back in the fall, Kay had figured out that the master programmer could run "loops" and reuse previously set cables and switches. They still celebrated that insight. But were there other secrets? Betty sat down again with the master programmer diagrams to puzzle things out.

In addition to loops, Betty realized that the master programmer could perform certain types of logic and, under certain circumstances, check to see if a specific condition was met. If the condition was met, then the program would proceed in one direction, and if not, then another set of steps would run.[9] Specifically, the master programmer was key to making the ENIAC general purpose and programmable, a flexibility needed for the trajectory calculations. If the Programmer wanted to determine if the result of a calculation was zero, indicating that the missile had hit the ground and finished its flight, she could do so. If the flight was finished, the Programmer could send the trajectory program into its final steps.[10] If not, the calculations could continue.

Much of this insight came from Betty's own analysis. "I learned the master programmer by myself." And she was impressed with its power. An IF statement is one of the most difficult things you can make a computer do, she recalled.[11]

It was heady stuff, and a Programmer would have to be pretty sophisticated to use the capability, but Betty took to it like a fly to honey. IF-THEN statements remain a fundamental part of programming logic today.[12]

Betty decided to talk with John Mauchly about the master programmer, and Jean joined her. As he started talking, John lit up and Betty and Jean realized that John was the creator of the master programmer. It was his favorite unit, his brainchild and "the soul of ENIAC."[13] In fact, it was the key to John's vision of ENIAC as the world's first general-purpose and programmable computer, in addition to being all electronic.

In Betty, John found a kindred spirit and complementary brilliance. He and Betty found camaraderie in their love of the master programmer and deep interest in general-purpose computing. Betty realized that with the power and versatility of programming, and with logic, she could program almost anything.

Jean, for her part, also enjoyed their talks with John. She found him a very responsive person. He listened carefully to women as well as men and followed what the person was doing. If Jean was singing a song, John would join and if she was quoting a line from *Alice in Wonderland*, "he'd finish the quotation for you."[14]

Bench Tests and Best Friends

During late November and early December of 1945, Kay and Fran continued to help run the Los Alamos project, and Betty and Jean made progress on their ballistics trajectory program. But in the midst of their work, Betty and Jean shared a burning question: How would they know if their trajectory program worked correctly?

Herman and John Holberton asked Marlyn and Ruth to find the answer. The two women would work with Betty and Jean to learn what incremental steps ENIAC followed to calculate a trajectory, and then calculate the answers by a separate method using the same steps.[1] Betty and Jean shared the steps of their pedaling sheets, and the group of four received the data for their test: the artillery and missile, temperature and crosswinds, humidity, distance to target, and other data they needed to run a sample trajectory for BRL. Both Marlyn and Ruth and Betty and Jean would run this test trajectory.

Marlyn and Ruth went off once again to find desktop calculators and a quiet space to work. The task was huge. Many additions and subtractions, multiplications, divisions and square roots had to be performed. It took a very long time, and the work was difficult and painstaking. But they were the right pair for a tough job. Ruth was a respected Computer known for her precise and excellent work, and Marlyn was a golden ring. In her thousands of hours with desktop calculations at the Moore School, Marlyn was known for never making a mistake. Even John Mauchly told stories of her remarkable work from their time on the radar project:

> [N]ever in all the time that she worked for him had she ever once made a mistake... She was just absolutely perfect, and he had never come across anybody like that. So that was Marlyn's qualification, she was a perfect Computer.[2]

Finally, it was done! Ruth and Marlyn knew what numbers should be stored in the accumulators, and printed to the IBM card punch, at each step of the trajectory program. When they finally got access to the ENIAC again, Betty and Jean would be able to stop the trajectory program at any step, examine the lights on top of an accumulator, and see if the number in temporary storage matched the number Marlyn and Ruth had calculated.

If yes, then the program was working and they could move on. If no, then they needed to stop and look for an error in the program's logic, wires, or switches. Even today, programmers run independent calculations to test their program in a technique often called a "bench test."

In the process of calculating the bench test, Marlyn and Ruth grew much closer. Their friendship had started in the Third-Floor Computing Team and grown in the Proving Ground. Now it deepened as, like Betty and Jean, Ruth and Marlyn spent almost as much time together outside of work as they did at the Moore School.

Marlyn, still fairly quiet and reserved, enjoyed being with her friend who was "very bubbly, very, very outspoken and a lot of fun to be with."[3] When Ruth moved to the Rebecca Gratz Club, offering housing for single Jewish women, Marlyn visited often. It was a wonderful place where everyone "gathered around the piano to sing, played cards in the lounge, and enjoyed the camaraderie typical of a sorority house."[4]

Sometimes they went out on the town. "We went to movies, we went to the concerts together, and plays when we could." Now that the war was over, even more events were taking place, and they had the time and money to enjoy them. When they were in a quieter mood, and Marlyn knew that Ruth had "a very quiet side," they went home to enjoy the warmth of Maryln's family, the Wescoffs.[5]

Marlyn's father had passed away, but her mother remained a warm entertainer. She was a wonderful cook who loved having people over and was delighted to host Ruth, who was so far from home. Ruth "came over a lot for dinner," Marlyn recalled.[6] Her aunt was often there and the house overflowed with music, traditional Eastern European foods of brisket and kugel, and great warmth.[7]

They also visited Ruth's parents a few times, heading by train up to New York City, then out to Long Island and their residence in Far Rockaway.[8] Known for its lovely, long beaches along the Atlantic, Far Rockaway was a playground for New Yorkers and a wonderful place for Ruth's parents to retire. A picture of the period, treasured by Marlyn for decades, shows the two young women on the beach in their bathing suits and robes with huge smiles on their faces. It felt good to be out in the sun, and together, the best of friends.

Parallel Programming

In the meantime, Betty and Jean continued to test their program. They "lived and breathed it." They could not have access to ENIAC itself, as it was still tied up with the Los Alamos problem, but they could track each other's logic and question each other's work. They worked through all the physical and logical details of the program, and it was an unusual working relationship:

> [We] each tried to find fault with what the other was doing. Instead of being angry when one partner found a fault, the other was delighted: it meant an error would not be left in the program.[1]

But there was one more snag. The program was too slow, and Jean and Betty knew that the speed of ENIAC's trajectory calculations would be key to its acceptance by BRL. Betty did not want to waste a moment.

What if ENIAC could be set up to run multiple steps at once? It seems a little crazy from today's perspective since the vast majority of computers for decades performed only one step at a time as "serial" or sequential processors, but ENIAC was a parallel processor.[2] It could run multiple operations at one time, but only if the Programmer was very careful and very smart.

The concept was straightforward: A program pulse could be fed into more than one unit at a time. Thus, an accumulator could start an addition and the high-speed multiplier could start a multiplication at the very same moment.

But timing was a problem. The accumulator ran ten times faster than the multiplier—it could perform 5,000 additions per second to the multiplier's 500 multiplications per second (still blindingly fast for its day). So if the next step of the program needed the results of the accumulator *and* multiplier, then the Programmer had to make sure that she waited long enough for the multiplier to finish its calculation before starting the next step with the accumulator and multiplier results.

Or as Jean would say with a deep belly laugh later, parallel programming of the ENIAC was hard because "you had to coordinate the timing... so that things could get back into synch."[3]

Betty embraced the challenge and knew that "we could have programmed the machine in serial, but we could not waste a single micro-second. And that's what made us do everything in parallel."[4] But thinking in parallel is difficult. Betty tried to figure out why and concluded:

> People don't think in parallel. They think in serial. You read a book in serial. You write in serial.... when you start having to do everything in parallel, you have to use another [type of] logic [to think] about what you're doing. You're collapsing everything in timewise.[5]

She later commented that programming ENIAC was "probably one of the hardest things I ever did."[6] Parallel programming to this day remains one of the most difficult challenges a programmer can face.

Betty and Jean set out to make as much of the ballistics trajectory program run in parallel as possible, and fortunately around this time, late winter 1945, Kay and Fran had come back to the ENIAC team. Nick and Stan had reached a rough answer for Los Alamos,[7] and although they would still keep their program on ENIAC for a little while, their need for a large team had dwindled.

Back with the ENIAC 6, Kay jumped in. She grasped the complexity of the parallel programming and its powerful contributions to the speed of the program, and she was able to help Betty and Jean with their complex work.[8] Together they finished tweaking their parallel trajectory program to fit onto ENIAC. "There were just little special ways in which you had to squeeze a little bit to get everything on at the same time," Kay said later.[9]

The six women were happy. Along the way, they asked questions of the engineers. No longer barred from the ENIAC room, they could enter whenever they wanted to look at a unit's control switches and plugs, or to ask an engineer a question. They were no longer exiles from the promised land.

Soon they would be part of the ENIAC team, and eating lunch and dinner with John, Pres, and the engineers. But the women still liked to be together, and whenever they had time, they went back to Lido's or sometimes to Arthur's Steakhouse, where they sat at a circular table, enjoying the steaks "carefully prepared and served just right," and each other's company.[10]

"I really loved working with those girls that were Programmers on the ENIAC with me," Kay said later with a warm smile.

The December holidays of 1945 were joyous and festive for the six women, and for the people of Philadelphia and across the country.

Most soldiers had returned home to be with their family and friends. For those still in Germany and Japan, the guns were quiet and the fighting was over.

On Christmas Day 1945, the headlines of the *Philadelphia Inquirer* announced the happy news of "the battleship *Washington*" arriving in New York Harbor with "1626 home-bound G.I.'s from Europe" and the sad news that General Patton had been laid to rest with military honors joining "the dead soldiers of his Third Army today beneath the thick, red clay of the Ardennes [Belgium], where they had fought together just a year before."[11]

But the holidays vanished quickly for the ENIAC 6, who were too wrapped up in their work to celebrate much. They continued to finish and improve their trajectory program well into the cold January of 1946.

And they kept asking themselves one big question: When will we be allowed to run our program on ENIAC?

Sines and Cosines

Outside the women's office, activity was buzzing. In January 1946, the Army decided to reveal the existence of ENIAC to the world. BRL had taken a big risk to sponsor technology that many said would not work. BRL had paid a total of $486,804.22 ($6.9 million today) and their gamble had succeeded brilliantly.[1] ENIAC was the world's first general-purpose, programmable, all-electronic computer and at least a thousand times faster than any other computer on Earth.[2] BRL wanted to take some credit and certainly wanted to celebrate, as did the University of Pennsylvania and Moore School too.

Herman worked closely with the War Department's Bureau of Public Relations. They set two dates for opening up the ENIAC room: February 1 for reporters with science and technology interests and February 15 for leaders of the US science and technology community.

In preparation for February 1, the Army sent invitations to the National Association of Science Writers and to science and popular magazine reporters. Then Herman, John, and Pres set out to write detailed sheets about the history of computing, advantages of general-purpose computers, a detailed description of ENIAC, and thoughts on how it might be used in the future for industrial and scientific purposes, as well as for military purposes. The press releases would serve as guides for the reporters when it came time to write their stories, and they needed detailed information.

Herman started his release, "Military Applications of ENIAC Described." John drafted "High Speed, General Purpose Computing Machines Needed," in which he predicted that with computers like ENIAC, in the future, "progress in mathematical physics and engineering will be greatly accelerated." Pres took the lead in writing "Physical Aspects, Operation of ENIAC are Described."[3] The men then edited each other's work and wrote some additional sheets.[4]

But their cooperation ended there. When it came time to write the release "Profiles of Personnel Who Developed ENIAC,"[5] Herman wrote himself in as a developer. John did not agree and rewrote the sheet, leaving Herman's bio for good measure.[6] In the end, the first person listed was Colonel Paul Gillon, now promoted, with a long biography, then Herman with a long biography. After that, Grist Brainerd was credited with handling Moore School administrative matters for the project. Finally, on page three, John's bio appeared as the man "whose original ideas for electronic general purpose digital calculators led to the development of the ENIAC." Deeply buried on page four was Pres's bio and description as "chief engineer of the ENIAC project" and "spark plug" of the team.[7]

Short mentions of Harold Pender, some of the ENIAC engineers, and Adele followed. But there was no mention of the six women. The names of Betty, Jean, Kay, Fran, Ruth, and Marlyn are nowhere to be found in the materials.

The War Department's Bureau of Public Relations staff took all of the essays and typed them neatly onto long legal sheets, preparing one set for each reporter. Each essay, however, had an Army warning at the top:

FOR RELEASE IN MORNING PAPERS, SATURDAY, FEBRUARY 16, 1946. FOR RADIO BROADCAST AFTER 7:00 P.M. EST, FEBRUARY 15, 1946.[8]

Although the reporters would see ENIAC on February 1, they would not be allowed to publish news stories about their visit until two weeks later. Clearly this "embargo" of the news by the Army was designed to create a big media splash to coincide with the February 15 demonstration. It would also give the reporters time to read the press releases carefully.

Press essays done, Herman moved on to design the activities for press day. He decided that he, Adele, and Arthur Burks would prepare a short series of programs showing off ENIAC's ability to do basic mathematics: addition, multiplication, squares and cubes, and sines and cosines.[9] They began to work.

In the meantime, an Army photographer arrived to take pictures of ENIAC to include with the press materials. Photographing the big computer was no easy task. Its big, black metal panels absorbed the light of a flash.[10] Special lights were set up. Plus, a ladder was brought in to allow the photographer to get a bird's-eye view of the ENIAC room and capture most of its great U formation.

And then John, Pres, Ruth, Jean, Herman, Fran, and other members of the ENIAC team were arranged around ENIAC and photos were snapped.

In a famous image, Pres and John stand in the middle of the room, Pres grinning as he pretends to turn a switch on a function table and John leaning against a central pillar. Jean stands in the back, on the far right, turning a switch on another function table, and Ruth is in the front, lower right, standing in front of some units facing the camera. Herman is positioned between them, in uniform, hands on a short cable. Private First Class Homer Spence stands in uniform in the back left.[11]

In a close-up showing only five units of ENIAC, Jean stands on the left and Fran stands on the right, their faces directed toward the computer.[12] And in another image, the photographer zooms in on Fran, feeding cards into the card reader and profiling her face, capturing woman and machine. Jean is on the far left of the image, but the camera is clearly focused on Fran.[13]

Each image is black and white and timeless. Produced in glossy images for the press kits and for the future ages to ponder. Although the people are dwarfed by the towering units, they are shown as professionals, calm and assured in front of the huge, new computer. Whether the photographer intended it or not, his images show the three groups whose work on ENIAC was invaluable: the inventors, the Army, and the women.

As a final preparation point, in late January, Herman notified the women that he would need them in the ENIAC room for the event. Not as Programmers, but as "hostesses."[14]

Friday, February 1, 1946, arrived and at 11:00 a.m., reporters from up and down the East Coast arrived at the Moore School. They represented the Philadelphia papers, the *New York Times*, and a range of scientific publications. They started with an introduction on a second-floor classroom of the Moore School and then made their way to the ENIAC room.

There they saw the ENIAC and the great U. John, Pres, Grist, Herman, and Major General Gladeon Barnes made some initial remarks.

Then Herman and Arthur began the presentation. They ran it just as they had planned: the addition and multiplication, a table of squares and cubes, and a table of sines and cosines.[15]

Then Herman announced he would run a part of a program he could not explain because it was classified (the Los Alamos program), and assured them that it was "an illustration of a long and complicated calculation."[16] The reporters came away a little confused.

A few people stayed to answer the reporters' questions after the event, and Ruth was asked to explain the functions of the units to any reporter who asked.[17]

Meanwhile, Betty and Kay headed over to the Franklin Institute to help with a lunch that had been arranged for the attendees.[18] The Franklin Institute, with a mission of promoting technology and innovation in Philadelphia, had agreed to provide the place.

Had Herman properly introduced Betty and Kay, the reporters would have known they were sitting with experts on the ENIAC and with individuals who could help them understand what they had just seen. Instead, he placed the women at the tables as hostesses to pour coffee for the male guests.[19] The two women were dismayed.

Betty sat next to a person from the publicity side of the Franklin Institute and at another table Kay sat next to the principal science writer for the *Philadelphia Bulletin*, "our big paper at the time," and near a provost for Penn. She learned that they left the ENIAC event rather lost and "[n]either of them had any concept of just what this meant at all."[20] Had the journalists only known who was pouring their coffee that day, one can only imagine the stories that would have run in their papers in the weeks to come.

After the reporters left, Herman, Adele, Arthur, and others realized that the event missed the mark, or was "rather dismal" as someone told Jean (who did not attend) later.[21] But this was only a trial run.

February 15 was the big demonstration. The one for the senior BRL officers and some of the country's most senior scientists and technologists. Herman needed something bigger and bolder to capture his audience's attention. And he had only two weeks to prepare.

A day or two later, Herman and Adele invited Betty and Jean to their apartment for tea.[22] This was odd because the couple never socialized with the Computers. Jean and Betty were happy to travel to their home and wondered about the purpose of their meeting.

Adele served tea in the living room, and a "big Persian cat... insisted on sitting on my lap," Betty remembered.[23] She was not crazy about cats but was far too interested in the conversation to shoo it away.

According to Betty, rather suddenly, Herman asked, "Could you put the trajectory on the ENIAC?" He added, "Is it ready to put on that?"

Betty and Jean said, "Of course."

He asked, "Well, could you have it ready for the demonstration?"

And the women said, "Of course, no problem."[24]

Betty and Jean looked at each other with excitement. "We had been dying to get our hands on the machine."[25] They could not wait to run their ballistics trajectory program.

Herman and Adele continued to ask questions: How complete was the trajectory? Could Jean and Betty set it up on ENIAC, debug it, and have it running for demonstration day?

The two women assured them they could.[26] They had checked their program many times by then.

"Okay, you'll be the demonstration problem," Herman directed.[27] They could start putting their program on ENIAC the very next day.

Betty and Jean left the apartment and stepped into the chilly February air. "We were thrilled," Jean said. "It was like a dream come true."[28]

The ENIAC Room Is Theirs!

The next morning, Betty, Jean, Marlyn, Ruth, Kay, and Fran marched into the ENIAC room, their heads high and their arms filled with pedaling sheets and bench tests. They were ready to start the most important job of their lives.

Their plan was simple: set up their trajectory program on ENIAC, run a sample trajectory, and compare its results to those on Marlyn and Ruth's bench test. Simple, but not very easy.

First, the women did an inventory of digit wires, program pulse cables, and other equipment they would need for the program. Ruth was "in charge of determining whether we had enough equipment... and we found that there weren't enough wires." One very long wire they needed was missing completely. "So Ruth had to get Clem [a wireman] to build it for us."[1] You could not go to the local hardware store to buy a digit wire or program pulse cable; each was specially made for ENIAC.

When the wires were laid out and ready to use, Ruth led the next step as well. As Nick and Stan had done with the Los Alamos problem, she prepared small cards with each switch and wire setting for ENIAC units.[2] She had made the cards with her beautiful handwriting, and the group slid them into the brackets on the front of most of ENIAC's units.

Then the women began the work of "setting up" the ENIAC—slinging wires and setting dials—except this time instead of Herman being the conductor, they ran their own show. Everyone worked together. "We had to hook the accumulators to the program trays to give you a sequencing," Betty said, remembering how fun it was:

> A cross between an architect and a construction engineer, because here you had all the parts, like the Lego parts or the Erector [set] parts . . . laid them all out in front of you.[3]

They followed the cards to string everything together, the accumulators, multiplier, and divider and square rooter, linking them together in the order of the trajectory program.

The women did the heavy work, too, slinging the fifty-pound digit trays (each eight feet long and two feet wide, encased in black metal and holding the equivalent of ten digit wires and designed to run across four units of ENIAC at a time). For Jean it felt a little like the bales of hay she used to sling with her father and brothers as she joined teammates in hefting a fifty-pound digit tray to the upper third of the ENIAC units where the digits ran.[4] Then she crouched low, with a partner, to place a fifty-pound program pulse tray along the bottom of the units, where the program pulses ran.

The women carefully followed Ruth's cards to set the dials of the divider and square rooter, for which number would be the numerator and which the denominator of a division, when and what numbers should be sent to the accumulator for addition and subtraction, and what results should be sent to the high-speed multiplier for additional computation.

And they set hundreds of switches on the function tables for the constants needed in the sample trajectory. They hoped no one would turn a switch when they were not watching.

It took about three days for the six women to set up ENIAC with their trajectory program, and they did it carefully. Then they moved on to the next stage of their work. Debugging their program. Was the logic right, and was each wire and switch set correctly?

Betty and Jean worked closely with Marlyn and Ruth to figure things out, one group holding the pedaling sheets and the other the bench tests.

They stood in the middle of the great U of ENIAC, holding a special tool John and Pres had created. A remote control black plastic box with four buttons connected with a long, thin black wire to the computer. Professor Brian Stuart of Drexel University's Computer Science Department, would later describe it as a "portable control station" and find that it had "a cable long enough to walk around the whole machine."[5] The perfect tool for debugging.

When the women clicked a particular button, it advanced the ENIAC "one ADD time."[6] This meant a program moved forward one addition, or one five thousandths of a second. With this remote control, they could control each incremental step of the great computer.

Each time they reached a calculation in the accumulator, or a number passed from another unit to the accumulator for processing, Betty and Jean stopped hitting the remote control and turned to look at the result in the temporary storage of the accumulator that had just been used. In the ten-by-ten grid of small lights at the top of the accumulator, they could see a few of the small bulbs lit at that moment, one for each digit of the number stored in the accumulator, 0 to 9. They read that number to Marlyn and Ruth, who checked it against the test results they had so carefully calculated.[7]

If the results matched, the women smiled and moved on to the next step of the program. If the results did not match, they stopped to recheck each switch setting, each digit wire and program pulse cable associated with all the units involved with that step of the program. They fixed anything they found that needed to be rewired or reset. They reran the program, ADD time by ADD time, until they got back to the step and checked the accumulator. When the results matched, they continued on. This process is now called *debugging*, and all programmers do it when checking their program.

Gradually, they made their way through the long trajectory program. At a certain point, they got so far into the program that it no longer made sense to go ADD time by ADD time. Now they thought about using the other button on the remote control, the button that ran the entire program, but that would move them to the end.

Then Betty got an idea. She marched over to the accumulator they were using at that latter step of the equation and pulled out the program pulse cable connecting that accumulator to the next step. She pulled it out.[8] Then Betty hit a different button on the portable control station, this time to run the entire program. At almost the speed of light, the ENIAC sped through all the steps of the program until it got to the step Betty wanted to see. Then it stopped.

With no program pulse to take the program to the next step, the calculation stopped, and the lights atop the accumulator where Betty had "broken" the pulse twinkled. Betty and Jean then read the results to Marlyn and Ruth and continued their debugging efforts.

Betty labeled her technique "breaking the point" and happily claims credit for the term still used today. "That's where the word break point comes from... We actually broke the point," she said with a laugh.[9]

The word *breakpoint* is still used in debugging today to stop a program and show the interim results at a step the programmer wants to see.[10]

Now the women had progressed through major portions of their program when they found something odd: Parts of their program already tested and debugged were no longer running correctly. It was odd because one day the step of the trajectory program worked and the next day it didn't. What could be going wrong?

And then they realized that the ENIAC room was not quite as secure as it could be. From time to time, they saw professors at the Moore School slipping in with guests to show off the huge computer and sometimes they turned switches or moved cables, although they knew perfectly well that they should not.[11]

Marlyn and Ruth became experts at checking the ENIAC for switch and cable settings. When they entered the room each day, they did a quick survey to see what might have been changed the night before after they left, particularly at eye and arm level.[12] They wished people without permission would keep their hands off ENIAC.

As the women moved toward the end of the trajectory program, another error began to pop up. A section of the program that worked the day before did not the next day, and Marlyn and Ruth assured them the switches and wires were properly set.

It took a little while, but they realized that the problem was not created by their program, or any of the switches or wires at the front of the ENIAC, but something else. A vacuum tube had blown, one of 18,000 in ENIAC, and if "one tube in one unit went wrong," Marlyn said, "we had to find it because nothing came out correctly; none of the calculations would be correct."[13]

But how could they find a blown vacuum tube? The ENIAC was huge, and plus, they were not allowed to touch the back of the units. Their domain was the front of the ENIAC, the electronics and wiring at the back of the ENIAC was the engineers' domain.

Instead, they focused on the diagnostics. The women ran their program ADD time by ADD time until they reached the point where a new error occurred.[14] That was likely the unit with the problem tube.

Initially, the engineers were a little reluctant to stop their own work to help, but when they realized the women could "debug the hardware down to a vacuum tube," they were "absolutely fascinated."[15] Out of 18,000, the women knew which one needed to be replaced. This was a win for the entire ENIAC team, and "once the engineers found that we could debug the ENIAC better than they could, they let us do it gladly," Jean recalled.[16]

In fact, the men came to rely upon the women to diagnose the vacuum tube issues, which we would call "hardware problems" today. In other words, the women had created a form of diagnostic programming.

Overall, the women's hard work, their ability to run a complex trajectory program "error-free," and "diagnose which component was making the error," won the acceptance of those around them.[17] "[T]he engineers generally treated us with a great deal of respect," Jean shared.[18]

And soon the respect became friendship. Working in the forty-by-sixty-foot room for days alongside the young engineers, they met some and learned more about others. Arthur was there; as was Chuan Chu, a young immigrant newly from China;[19] Kite Sharpless, in his late thirties, a Moore School graduate with a gentle sense of humor;[20] and Jack Davis with his love of jokes. Bob Shaw, who Jean and Betty already knew, met the whole team and everyone liked him. "A true Renaissance man... a great engineer, a good writer, a good talker."[21] He had a wild sense of humor and, as an albino, such poor eyesight that he kept his face close to the hot vacuum tubes and everyone thought he would catch himself or his papers on fire.[22] John and Pres

Photographs of the ENIAC 6 taken in the 1940s and early 1950s

Jean Jennings (Bartik), 1940s. *Courtesy of the Bartik Family*

Kathleen McNulty (Mauchly Antonelli), 1940s. *Courtesy of Bill Mauchly*

Frances Elizabeth "Betty" Snyder (Holberton), 1940s. *Courtesy of Priscilla Holberton*

Frances Bilas (Spence), 1947. *Courtesy of the Spence Family*

Marlyn Wescoff (Meltzer), Temple University Graduation, June 1942. *Courtesy of the Meltzer Family*

Ruth Lichterman (Teitelbaum), early 1950s. *Courtesy of the Teitelbaum Family*

Computer Teams at the Moore School 1942–45

The Third-Floor Computing Team of the Army's Philadelphia Computing Section located at 3436 Walnut Street. The women worked as a team for years: two weeks on days and two weeks on nights calculating ballistics trajectories using electromechanical desktop calculators. Alyce Hall at far left and Marlyn Wescoff at far right. (Ruth would join the team later.) *Courtesy of the Meltzer Family*

Kathleen McNulty (Mauchly Antonelli) supervises ballistic trajectory calculations on the differential analyzer in the basement of the Moore School during WWII. She, far left, reads and records the output of the calculations as Sis Stump, far right, inputs data. Alice Snyder, middle, checks shafts and gears. *University Archives and Records Center, University of Pennsylvania*

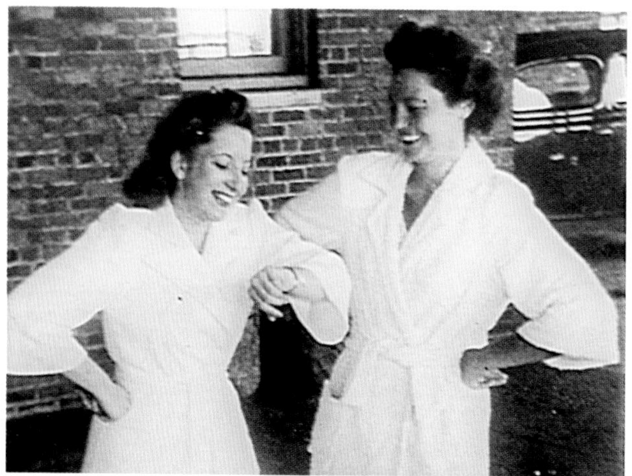

Marlyn Wescoff (Meltzer) and Ruth Lichterman (Teitelbaum) enjoy a rare day off at the beach, probably while visiting Ruth's parents in Far Rockaway, NY. *Courtesy of the Meltzer Family*

SHARPLY POINTED PROJECTILE
MOVING THROUGH THE AIR

$$(a^2 - u^2)\frac{\partial u}{\partial x} - uv\left(\frac{\partial u}{\partial y} + \frac{\partial v}{\partial x}\right) + (a^2 - v^2)\frac{\partial v}{\partial y} + \frac{a^2 v}{y} = 0,$$

$$\frac{\partial v}{\partial x} - \frac{\partial u}{\partial y} = 0$$

A typical equation used by the Computers for calculating a ballistics trajectory. It took into account various weather conditions on the battlefield and the type of artillery gun and shell being used. *National Archives*

The Moore School of Electrical Engineering at the University of Pennsylvania, circa 1929. Construction of the third floor began late in WWII. *University Archives and Records Center, University of Pennsylvania*

Women programming and debugging ENIAC at the Moore School, 1945–46

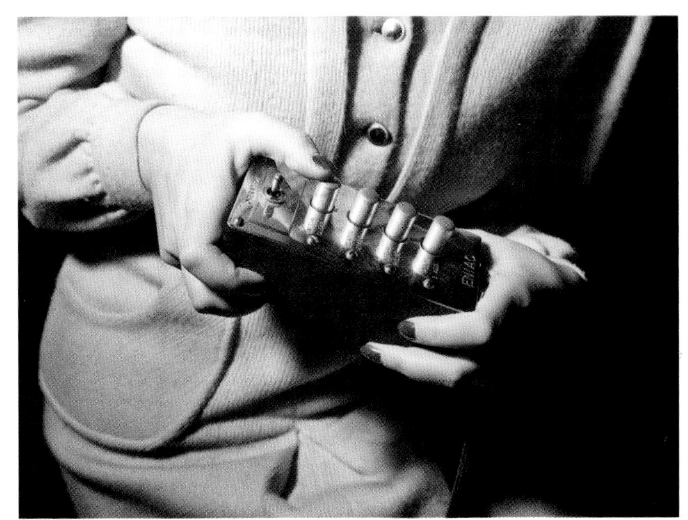

The portable remote control tester was a critical tool used by the ENIAC 6 in debugging their ballistics trajectory program. It allowed them to advance step-by-step through their program and helped them to find problems in logic and hardware. © *Bettmann / Getty Images*

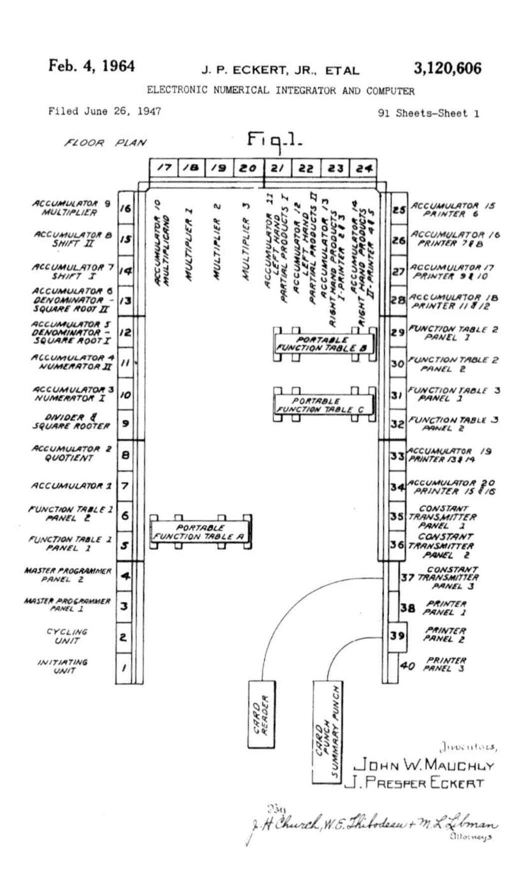

The floor plan of the units of ENIAC in their U-shape configuration at the Moore School in 1945–46. This floor plan was included in John W. Mauchly and J. Presper Eckert's patent application, US Patent 3,120,606, filed June 26, 1947.

Ruth Lichterman (Teitelbaum), left, and Marlyn Wescoff (Meltzer), right, feed punched cards into the IBM card reader to provide data as input to their program on ENIAC. *Special Collections Research Center, Temple University Libraries, Philadelphia, PA.*

For decades, this famous photograph of two ENIAC Programmers was shared without the women being named. Jean Jennings (Bartik) stands on the left and Frances Bilas (Spence) on the right. Part of the photographs taken in preparation for Demonstration Day. *University Archives and Records Center, University of Pennsylvania.*

This famous image of the ENIAC team was taken by an Army photographer for Demonstration Day, shared with reporters, and published in newspapers around the country. For over fifty years, the women were not named in the caption. Left to right: Pfc. Homer Spence, J. Presper Eckert, Dr. John Mauchly, Jean Jennings (Bartik), Capt. Herman Goldstine, and Ruth Lichterman (Teitelbaum). *University Archives and Records Center, University of Pennsylvania*

The traditional way the ENIAC story was told for over fifty years—all-male, with representatives of the Army, Moore School, and inventors of ENIAC. Left to right: J. Presper Eckert, ENIAC Co-Inventor; Dr. John Brainerd, Dean of Research, Moore School; Sam Feltman, Chief Engineer for Ballistics, Army Ordnance Dept.; Capt. Herman Goldstine, BRL Liaison Officer to Moore School; Dr. John Mauchly, ENIAC Co-Inventor; Dr. Harold Pender, Dean of Moore School; Maj. Gen. Gladeon Barnes, Chief of Army Ordnance Research and Development Service (AORDS); and Col. Paul Gillon, Chief, Research Branch of AORDS. *University Archives and Records Center, University of Pennsylvania*

The ENIAC team bonded over their work on ENIAC and lunches and dinners together, where they brainstormed about the future of computing. The Lido Restaurant was a favorite destination. February 1946, left to right: Pfc. Irwin Goldstein, Frances Bilas (Spence), Homer Spence, James Cummings, Marlyn Wescoff (Meltzer), John W. Mauchly, Ruth Lichterman (Teitelbaum), Jean Jennings (Bartik), and Kathleen McNulty (Mauchly Antonelli). *Courtesy of Bill Mauchly*

On Jean's wedding day, December 14, 1946, the ENIAC team came together to celebrate. Left to right: Frances Elizabeth "Betty" Snyder (Holberton), maid of honor; William "Bill" Bartik, groom; Jean Jennings Bartik, bride; and Dr. John Mauchly, who walked Jean down the aisle. *Courtesy of the Bartik Family*

ENIAC continues operations at Aberdeen Proving Ground, Aberdeen, Maryland

In 1947, ENIAC began operations at the Ballistics Research Laboratory on Aberdeen Proving Ground. Ruth Lichterman (Teitelbaum), lower right, continued as a Programmer and supervisor, and Ester Gerston, center, a WWII Computer, continued her work for the Army as a second-generation ENIAC Programmer. *© Corbis Historical / Getty Images*

Dozens of women and men would go on to program ENIAC at Aberdeen for important military, academic, and commercial problems until 1955. This famous image shows second-generation Programmers Gloria Gordon Bolotsky, crouching, and Ester Gerston programming ENIAC with thick digit cables, thin program pulse wires, and switches. *US Army Research Laboratory*

Anniversaries and Reunions

In 1985, Kay McNulty Mauchly invited all the ENIAC Programmers to Philadelphia for a reunion. It was the last time they would be with Ruth. Back row, left to right: Ruth Lichterman Teitelbaum, Adolph Teitelbaum, Frances Bilas Spence, Homer Spence, Marlyn Wescoff Meltzer, and Kathleen McNulty Mauchly. Front row, left to right: Betty Snyder Holberton and Jean Jennings Bartik. *Courtesy of the Bartik Family*

A special reception for the ENIAC Programmers at the fiftieth anniversary of ENIAC in 1996. Left to right: Kathy Kleiman (author), Jean Jennings Bartik, Marlyn Wescoff Meltzer, and Kathleen McNulty Mauchly Antonelli. Betty Snyder Holberton is in front. *© Steven M. Falk. Courtesy of First Byte Productions, LLC*

Kathleen McNulty Mauchly Antonelli, left, and Jean Jennings Bartik, right, hold an ENIAC decade counter with vacuum tubes at the Moore School Building of Penn Engineering in 2001. This photograph was taken in April 2001 for a cover story profiling Jean, class of 1945, in *Northwest Alumni Magazine* for Northwest Missouri State University. *Northwest Missouri State University Archives*

continued to monitor night and day to help with testing as everyone headed into the big event.

The whole ENIAC team, women and men, began spending time together. Sometimes, if they were working late, they would slip away for dinner. This included Fran, Kay, Marlyn, Ruth, Jean, Betty, some of the engineers, and, many times, John. They would order dinner and as they waited, talk about computers.

As Jean remembers, they asked each other, "How could they be used? How were we doing with the ENIAC? What was the future going to be like?"[23]

Like technologists everywhere, their napkins became the diagrams on which they sketched their visions of computing to come, a future they were helping to create.

The Last Bugs Before Demonstration Day

Finally, the program was working, running and producing results that corresponded to the bench test. Their program was a success, and it ran fast, about twenty seconds to calculate a trajectory that had taken them thirty to forty hours to calculate by hand. Their parallel programming had worked; they had maximized every microsecond of the program.

On the morning of Valentine's Day, one day before Demonstration Day, Betty and Jean were puzzling over the last bugs. There were only two left. One bug was the printouts not coming out cleanly. Instead of printing the points of the trajectory as whole numbers, they came out as decimals: .8, 1.8, 2.8, and 3.8, instead of 1, 2, 3, and 4.[1] This error troubled Betty.

In the trial run at 11:00 a.m., with a few visitors in the room, Betty and Jean ran the trajectory and Ruth took the punched cards over to the tabulator, ran off printouts, and gave them to some of the visitors. "Why is this printing out .8?" one person asked, and Betty sighed. This was a bug they would have to fix before tomorrow.

As for the second bug, when Jean looked at the printout, she noticed that the trajectory "dug a little hole."[2] For some reason, the program did not stop when the missile hit its target or the ground; it went just a little bit farther. But why?

Jean and Betty set to work trying to solve both bugs. Demonstration Day was tomorrow! They worked through the details again and again, oblivious to everything but the ENIAC and their pedaling sheets.

They were surprised when Harold came in and asked how they were doing. They said they were fine, but he could see how hard they were working. In his hand, "he had a little brown bag." After chatting with them for a bit, "he laid the bag down on the table that we were working on and walked out of the room."

"Go to it," he said as he left,[3] and "keep up the good work."[4]

They opened the bag. Inside was "a fifth of liquor," and Jean and Betty let out "a big howl."[5] Other than an occasional cocktail, they did not drink hard alcohol.

It was odd, but just a few days earlier, John Mauchly had dropped by with a similar gift. He had brought a bottle of apricot brandy and given both women "a small glass" to drink. It was "the first time I had ever tasted it," Jean recalled, "and it was delicious."

The two acts meant a lot to the women: "Like Mauchly's gesture, Dean Pender's gesture impressed us and made us understand how much he wanted us to succeed, and how much this event meant to the University of Pennsylvania."[6]

With that new appreciation of their important role in the event to come, the women kept working. They solved the printout problem, but late that night, they had not yet solved the small hole still being dug by the trajectory. They stayed late, until about midnight, and then Betty had to catch the last train to Narberth. "[W]e shut off the lights and went home," Jean recalled, "thinking that we would just have to let it go."[7]

But the next morning they arrived early, about 7:45 a.m., now dressed in their best professional suits for Demonstration Day. When Jean looked at Betty, she knew something had happened. Betty had a determined look in her eye and a new confidence in her bearing.

Jean watched as Betty marched to one of the many switches of the master programmer, turned it one notch, and announced that their program was finished. Jean was amazed.

In her sleep, Betty had gone through all of ENIAC's switches and wires and knew exactly "which switch of the three thousand on the ENIAC to reset" and exactly what position it should be.[8] For years to come, Jean would declare that Betty solved more programming problems in her sleep than most people solve when they were awake.[9]

Betty, for her part, was more modest. She had forgotten that zero was a number that represented ground on the master programmer.[10] By remembering that, she realized that she needed to reset one of the switches on the master programmer from 1 to 0 to make the loop end in the right place. It was "my first Do-Loop error," she admitted,[11] and nearly every Programmer to follow would have a similar experience. Afterward, Betty always carefully thought about mistakes Programmers might easily make with the tools they were using.

The women reran the trajectory program and "[w]e were ecstatic. The program worked perfectly and we were ready to go."[12] They took the final punched card deck of the now-perfect trajectory and handed it to Kay, Marlyn, Ruth, and Fran. Together the group would create printouts on the accordion-style paper of the tabulator for each guest to take home, a trajectory souvenir at the end of the event.[13]

Then the six women looked at each other and began their final preparations for the ENIAC demonstration. It was scheduled for 11:00 a.m.

Demonstration Day, February 15, 1946

The Moore School stood ready as people began to arrive by train and trolley. John and Pres, as well as the engineers and deans and professors of the university, wore their best suits and Army officers were in dress uniform with their medals gleaming. The six women wore their best professional skirt suits and dresses.

Kay and Fran manned the front door of the Moore School. As the scientists and technologists arrived, some from as far as Boston, the two women welcomed them warmly. They asked everyone to hang up their heavy winter coats on the portable coat racks that Moore School staff had left nearby. Then they directed them down the hall and around the corner to the ENIAC room.

Just before 11:00 a.m., Fran and Kay ran back to be in the ENIAC room when the demonstration began.

As they slid into the back of the room, everything was at the ready. At the front of the great ENIAC U, there was space for some speakers, a few rows of chairs, and plenty of standing room for invited guests and ENIAC team members. Across the room, Marlyn, Betty, and Jean stood in the back and the women smiled to each other. Their big moment was about to begin. Ruth stayed outside, pointing late arrivals in the right direction.

The room was packed and was filled with an air of anticipation and wonder as people saw ENIAC for the first time.

Demonstration Day started with a few introductions. Major General Barnes started with the BRL officers and Moore School deans and then presented John and Pres as the co-inventors. Then Arthur came to the front of the room and introduced himself as the master of ceremonies for the ENIAC events. He would run five programs, all using the remote control box he held in his hand.

The first program was an addition. Arthur hit one of the buttons and the ENIAC whirled to life. Then he ran a multiplication. His expert audience knew that ENIAC was calculating it many times faster than any other machine in the world. Then he ran the table of squares and cubes, and then sines and cosines. So far, Demonstration Day was the same as the one two weeks earlier, and for this sophisticated audience, the presentation was pretty boring.

But Arthur was just getting started and the drama was about to begin. He told them that now he would run a ballistics trajectory three times on ENIAC.

He pushed the button and ran it once. The trajectory "ran beautifully," Betty remembered. Then Arthur ran it again, a version of the trajectory without the punched cards printing, and it ran much faster. Punched cards actually slowed things down a little bit.

Then Arthur pointed everyone to the grids of tiny lights at the top of the accumulators and urged his attendees to look closely at them in the moments to come. He nodded to Pres, who stood against the wall, and suddenly Pres turned off the lights. In the black room, only a few small status lights were lit on the units of ENIAC. Everything else was in darkness.

With a click of the button, Arthur brought the ENIAC to life. For a dazzling twenty seconds, the ENIAC lit up. Those watching the accumulators closely saw the 100 tiny lights twinkle as they moved in a flash, first going up as the missile ascended to the sky, and then going down as it sped back to earth, the lights forever changing and twinkling. Those twenty seconds seemed at once an eternity and instantaneous.

Then the ENIAC finished, and darkness filled the room again. Arthur and Pres waited a moment, and then Pres turned on the lights and Arthur announced dramatically that ENIAC had just completed a trajectory faster than it would take a missile to leave the muzzle of artillery and hit its target. "Everybody gasped."[1]

Less than twenty seconds. This audience of scientists, technologists, engineers, and mathematicians knew how many hours it took to calculate a differential calculus equation by hand. They knew that ENIAC had calculated the work of a week in fewer than two dozen seconds. They knew the world had changed.

Climax complete, everyone in the room was beaming. The Army officers knew their risk had paid off. The ENIAC engineers knew their hardware was a success. The Moore School deans knew they no longer had to be worried about being embarrassed. And the ENIAC Programmers knew that their trajectory had worked perfectly. Years of work, effort, ingenuity, and creativity had come together in twenty seconds of pure innovation.

Some would later call this moment the birth of the "Electronic Computing Revolution."[2] Others would soon call it the birth of the Information Age. After those precious twenty seconds, no one would give a second look to the great Mark I electromechanical computer or the differential analyzer. After Demonstration Day, the country was on a clear path to general-purpose, programmable, all-electronic computing. There was no other direction. There was no other future.

John, Pres, Herman, and some of the engineers fielded questions from the guests, and then the formal session finished. But no one wanted to leave. Attendees surrounded John and Pres, Arthur and Harold.[3]

The women circulated. They had taken turns running punched cards through the tabulator and had stacks of trajectory printouts to share.[4] They divided up the sheets and moved around the room to hand them out. Attendees were happy to receive a trajectory, a souvenir of the great moment they had just witnessed.

But no attendee congratulated the women. Because no guest knew what they had done. In the midst of the announcements and the introductions of Army officers, Moore School deans, and ENIAC inventors, the Programmers had been left out. "None of us girls were ever introduced as any part of it" that day, Kay noted later.[5]

Since no one had thought to name the six young women who programmed the ballistics trajectory, the audience did not know of their work: thousands of hours spent learning the units of ENIAC, studying its "direct programming" method, breaking down the ballistics trajectory into discrete steps, writing the detailed pedaling sheets for the trajectory program, setting up their program on ENIAC, and learning ENIAC "down to a vacuum tube."[6] Later, Jean said, they "did receive a lot of compliments" from the ENIAC team,[7] but at that moment they were unknown to the guests in the room.

And at that moment, it did not matter. They cared about the success of ENIAC and their team, and they knew they had played a role, a critical role, in the success of the day. This was a day that would go down in history, and they had been there and played an invaluable part.

As guests began to leave the ENIAC room, Fran and Kay ran back to the front of the school. They did everything in reverse, helping men find their coats and hats, gloves and scarves. It was still a cold February day in Philadelphia, and everyone would need to be bundled up.

After the guests left, the six women regrouped. They had a lot to talk about. The events of the day and the famous mathematicians and scientists they had met.

They noted the face-lift John and Pres had given ENIAC the night before—cutting white Ping-Pong balls in half to put them over the light tips and painting numbers on them so everyone could read them. Apparently it had taken them half the night, but the effect was spectacular as the tiny lights danced and lit up the dark room like the "marquees in Las Vegas."[8]

They talked about the demonstration and its success, and the oddity of their not being introduced. The women had different reactions.

Jean and Betty saw chauvinism in the event. Although the work had been done by women and men, Jean felt the presentation became "a men's show,"[9] and Betty felt the same way. In not being introduced that day, she said, "We expected that. In those days the women were not recognized at all. So it was just a normal thing."[10]

Marlyn knew that, without being told of all their thousands of hours of underlying work, the attendees would think they were "the operators of that machine and that was it."[11]

But Kay could see how the introductions made sense even if she did not like being overlooked. After years of fighting over funding, delays in ENIAC's delivery, and other problems, now was a time for the men to come together:

> I think it was the Moore School's day. In other words, they were honoring Eckert and Mauchly and the engineers who

> had built the machine. And they were honoring the brass of Aberdeen that had the gumption, you might say, to put up the money for such a speculator type of thing.[12]

As for the women, she noted, "We women sort of fell between the cracks because we didn't belong to the Moore School, and we were just Programmers. We were just Computers as far as the [Army] brass was concerned."[13] Just Programmers.

But for Kay, that didn't take away for a second from how valuable and unique she felt the women's contributions were to the day:

> In retrospect, I think that we were like fighter pilots. I mean here was this great, great machine. But you couldn't just take any ordinary pilot and stick him into a fighter pilot [plane] and say, "Go to it now, man!" I mean, that was not the way it was going to be.[14]

She knew they had done something only a few people in the world had been trained to do, and that only a few people could do. They all knew.

They spent more time talking, and more time putting things away, and then "it was the end of the day and I went home," Marlyn recalled. There was an event to follow that night, but "I wasn't invited to anything."[15] They all said good night as the sun set early in the cold Philadelphia winter night and went off.

"On probably no other day of my life have I experienced such thrilling highs and such depressing lows," Jean shared. "Betty felt the same way."[16] They trudged home, very tired, in the bitter cold "feeling let down after all the excitement."[17] Soon it was time to part ways, Jean for her trolley and Betty for her train out to the suburbs. Each woman left absorbed in her own thoughts.

That night, at Penn's Houston Hall, the student union, in the grand banquet room,[18] Penn hosted a great dinner for the Army, deans, and invited guests of the day to celebrate the "unlimited scientific future of the newest technology development"—ENIAC.[19] The guest of honor was Dr. Frank Jewett, director of Bell Labs and President of the National Academy of Sciences. Herman had wanted President Eisenhower, but he settled for Jewett.

Jewett didn't know much about ENIAC but noted it was "a tool for noteworthy scientific advancement."[20] The President of Penn himself, Dr. George McClelland, joined Jewett in leading the evening and crediting his own university with foresight.

Major General Barnes congratulated John and Pres on their achievements, and noted the Army's role in funding it. ENIAC, he said, would soon be stationed at the Proving Ground "for the solution of many problems which may be presented to us by the scientists of the country."[21] BRL already had a plan to use ENIAC for both military and nonmilitary problems.

Before the women left that night, they knew about the big dinner and that not a single woman, not even Adele, had been invited. They knew also that initially the young engineers of ENIAC had not been invited, but John and Pres had put their foot down. If the dinner was to honor ENIAC, it had to include the engineers who built it.

But no one, not even their supervisors Herman or John Holberton, put their foot down to insist the young women who programmed its trajectory needed to be invited too. And "being ignored was hurtful," Jean admitted.[22]

So that night, the men ate "Bisque of Lobster" and "Filet Mignon Au Jus or Broiled Salmon Steak,"[23] toasted each other with wine and cognac and smoked cigars. After years of restraint during the rationing of the war, it was the time to celebrate their victories with a surfeit of food, alcohol, and backslapping. Army and academia had come together to help win the war, and now they found that they had sponsored the creation of a new technology likely to change the postwar world. It was a night of manly celebration.

The women had their own celebrations, albeit smaller ones. On February 15, the Army allowed publication of the ENIAC story and the press responded with enthusiasm. The *Philadelphia Record*'s front page proclaimed amusingly, BLINKIN' ENIAC'S A BLINKIN' WHIZ, with a subtitle announcing, ELECTRONIC CALCULATOR OPERATING AT PENN DOES WORK OF 20,000 PERSONS.[24] The picture of Jean and Fran, with Private First Class Irwin Goldstein and ENIAC, was published for all to see, just as Army's Bureau of Public Relations had hoped, with a caption that did not name a single person in the picture.

The secrecy of their work as Computers and Programmers was finally over, and Marlyn, Fran, Kay, Ruth, Betty, and Jean were happy they could finally tell their families, roommates, and friends what they had done during the war, and for so many months afterward. And they did.

Marlyn's newspaper-reading family "saw all the publicity and they were interested." They knew a bit of what she was doing "but they didn't know what kind of information was coming off."[25] She happily told them the rest of the story.

The events of the day became exciting dinner conversation across the city as other groups discussed the newspaper stories, listened to the radio broadcasts, and shared the news of ENIAC. The ENIAC Programmers shared the day, and the success of their trajectory, with their families and friends. Each woman was justifiably proud of her accomplishments, as well as those of the other women and the entire ENIAC team.

In Betty's family, there was a special knowledge and appreciation. For her family "was so much ahead of our time for two generations," with her grandfather on the US Electrical Commission to bring electricity around the country and a founder of the National Bureau of Standards.[26]

That night, Betty shared with them her work and the great ENIAC she had programmed. A third generation of scientists and technologists was in the Snyder family, innovation now continuing through their daughters as well as their sons.

Betty added some of her thoughts about a future filled with general-purpose, programmable computers running at almost the speed of light. She wanted to stay involved in this technological "volcano" that was just starting to erupt.[27]

A Strange Afterparty

That day and in the days to come, the ENIAC made headlines up and down the East Coast, and then across the country. Friends and family cut and sent stories to Marlyn:

> *New York Times*, Electronic Computer Flashes Answers, May Speed Engineering, February 15, 1946
>
> *Boston Daily Globe*, World's Fastest Calculator Cuts Years' Task to Hours, February 15, 1946
>
> *Chicago Daily Tribune*, Robot Calculator Knocks Out Figures Like Chain Lightning, February 15, 1946[1]

She saved them for the rest of her life. Betty cut out the articles, too, and started a scrapbook called *We Were There When It Happened.*[2]

The *Boston Globe* article even pictured Marlyn and Ruth and captioned it "Feeding Robot's Reader—Ruth Lechterman (sic) (left) and Marlyn Wescoff operate the reader of the new computer." They posed with punched cards as if ready to feed them in.[3]

The *Stanberry Herald-Headlight* ran a story about Jean's accomplishments as one of six women "who program the solution of various mathematical problems for the machine and also help in the operation of the machine." In bragging hometown fashion, the reporter wrote that it was "no great surprise to know that she is holding such an important position" given her outstanding success in high school.[4] Jean had at last done something no one in her family, or her town, had done, and everyone now knew it.

But for the most part, the photograph captions named John, Pres, Herman, and Major General Barnes and otherwise did not name individuals. It was the huge, black ENIAC that the papers wanted to show off.

Betty lamented that the *New York Times* article did not describe the trajectory that had so captured the attention of the audience on February 15,[5] but the article was written based on the earlier February 1 press event and the technical press releases written by John, Pres, and Herman. The reporter never saw the powerful trajectory demonstration that took place two weeks later.

After Demonstration Day, the doors of the ENIAC room stayed wide open and many people came to see ENIAC, including more reporters, *Movietone News* producers, scientists, and educators. Kay remembers the film crews recording.[6] Soon a *Movietone News* clip appeared in theaters. In it, a booming voice asks, "Are people becoming obsolete?" and shows Kay bringing the accordion-style printouts to John and Pres.[7]

The big voice continues: "A giant electronic brain has started cogitating at the University of Pennsylvania. It's made of vacuum tubes like your radio and it can add up a column of figures a yard long in a second. It's the world's first electronic computer. Right now it's solving mathematical problems for the U.S. Army, but who knows? One day a machine like this may check up on your income tax."[8]

The audience, of course, laughed. The idea that such a giant computer would one day be part of their lives was preposterous. Giant computers were intended for giant problems, military-style problems.

But Jean did not laugh. She was concerned that people might believe that ENIAC was actually thinking when it certainly was not: "The ENIAC wasn't a brain in any sense; it couldn't reason, as computers still cannot reason, but it could give people more data to use in reasoning."[9]

Overall, the publicity was good and ENIAC received attention in stories over the next few months from coast to coast and up into Canada.

It should have been a time of glory, excitement, celebration, and perhaps even a little relaxation. But Dean Irven Travis had other plans. Irven had been away during the war, working for the Navy, and when he returned, he was "placed in command of all the Moore School research."[10] He took over some of the functions that Grist had been doing, and he wasn't very happy with what he found.

Irven wanted the Moore School to own all of the patents of inventions taking place under its roof, including the patents for ENIAC and its next-generation successor, EDVAC (Electronic Discrete Variable Automatic Computer). It didn't matter that the contracts for these computers had been negotiated and signed years earlier, ENIAC in 1943 and EDVAC in 1944. It did not matter that the Moore School at the time did not care about or want the patents, but that the Army did want to protect the inventions and had encouraged John and Pres to apply for appropriate patents, both to protect the Army from future patent trolls and perhaps even to create a billion-dollar industry.

None of this mattered to Irven, including that John and Pres had a deal "signed by the president of Penn" granting them the patents.[11]

Irven thought he could bully his way into owning them, and on Friday, March 15, 1946, a month after Demonstration Day, he announced in a staff meeting that "he needed a patent release from all his people."[12]

John, Pres, and ENIAC engineers Bob Shaw, Jack Davis, and some others did not agree. "Even Carl Chambers, the supervisor of research and no radical, thought this was out of line."[13]

Irven offered them more time to consider. A week later, on Friday, March 22, John and Pres received a letter "written by Travis and signed by Pender" demanding that they turn over their patent rights and agree to stay at the Moore School for two more years.[14] They had until 5:00 p.m. that day.

At 5:00 p.m., John and Pres "delivered identical letters" to the Moore School tendering their resignations.[15] They would keep their patents but leave their jobs. That day they left the ENIAC and their successor project EDVAC, which was to be the world's first "stored-program" computer. It was the basis of modern computer architecture. Without their leadership and brilliance, EDVAC would be delivered years late to BRL.

The women were stunned. "Everyone was flabbergasted and thought Travis was crazy," Jean said.[16]

And those assessing the history would say that it was a fatal error for the Moore School. Joel Shurkin, a *Philadelphia Inquirer* reporter for many years and author of *Engines of the Mind: The Evolution of the Computer from Mainframes to Microprocessors* in 1984 wrote of this pivotal moment:

> Although few universities grant their employees complete commercial rights, Penn seemed to be alone in demanding that their scientists give up all such rights. In one simple act, Travis destroyed the University of Pennsylvania's substantial

lead in computer sciences and probably cost it millions of dollars in potential licensing fees and royalties, and unmeasurable amounts of prestige. The university never recovered from the firing. In the history of American higher education, Travis's decision surely stands in a class by itself. At the Moore School, the ENIAC firings are to this day called the "great-might-have-been."[17]

Now the doors of the ENIAC room were slammed closed to the two men who had invented the idea of ENIAC, then lived and breathed it for three years as they oversaw the team that designed, built, and brought it to life.

Hundred-Year Problems and Programmers Needed

John and Pres left, and the ENIAC 6 remained at the Moore School with full and complete access to the ENIAC room. Now, over seven months since the war had ended, the Army had no intention of sending these women home.

For BRL wanted to know what it had purchased. What was a general-purpose, programmable, all-electronic computer, and how much could it do? To answer these questions, BRL "agreed to provide time free of charge" to a few world-class mathematicians and scientists, and the Moore School agreed to host them.[1] In spring 1946, a half-dozen scientists and technologists came from across the United States and from the United Kingdom to use ENIAC.

Initially, there was a sense that the mathematicians and scientists would learn to program ENIAC themselves, each taking the time to learn the units of ENIAC and its special "direct programming" technique. But the mathematicians who arrived knew better. The ENIAC was huge and difficult to use. They didn't want to take the time to learn how to program ENIAC; they wanted to find Programmers who could understand their equations and let them do the work. They quickly glommed on to the six women Programmers whom BRL was wise enough to keep in its employ. The ENIAC 6 were, of course, happy to help.

The problems they brought were called "hundred-year problems" because, as Kay explained, "it would take a hundred years using the methods that people had then, pencil and paper and maybe a desk calculator, to actually arrive at the solution to this problem."[2] They were basically unsolvable problems for the time, but that did not seem to deter the mathematicians, or their new Programmers.

Early to arrive was Dr. Douglas Hartree in April 1947, from the United Kingdom. He came with a big problem and a big appetite. After many years of serious rationing, he was thrilled to be enjoying the bounty of US farms. His big question involved the study of turbulence around the wing of a plane. Kay was assigned to help him calculate airflow around an airplane wing. This was a question of great interest at the time to military and civilian aviation.

Douglas brought some preliminary programming ideas and handed them over to Kay, who took them from there:

> He had already gotten some information about how to program before he came. I mean he had the general idea. But he didn't know a lot of specifics. So what I did was that I went over his program with him and helped him. He had already broken it down from the differential equations into the numerical equations that had to be solved. And so I went through with him and checked that. And then together we set up the machine.[3]

They plugged the wires, set the switches, and then tested and debugged the program. Then they ran it for several types of wing shapes. Kay enjoyed their work together, finding Douglas a complete delight to work with in every way. He was "interesting, fun, thoughtful."[4] Sometimes Douglas would "go away for days and leave me there running his problem," and Kay was happy to do so. He traveled to the University of Michigan and University of Wisconsin to visit colleagues, sending friendly letters back to Kay, thanking her for her work and feeling confident that his program was in good hands.[5]

Kay stayed and ran the program, carefully recording the results.

John Goff, dean of the Towne Scientific School next door, brought his question about the "thermodynamical properties of gases," how various gases respond to different temperatures and pressures. This was another question of great interest to US industry and military in the late 1940s.[6] Goff quickly found Ruth and Marlyn.

The two women helped him break down the calculus equations into small steps ENIAC could handle, and then they created pedaling sheets and set up ENIAC for the problem. They ran it again and again with data for different gases, temperatures, and pressures. Computing power solving another previously unsolvable problem.

Betty worked with Hans Rademacher, from the Penn mathematics department, to investigate how ENIAC "rounded off" numbers for the calculations it was doing. Computers, like people, need to set a precision for their calculations. While we may say that we want to give someone a third of dollar, we cannot give them the precise amount, as a third of a dollar is 33.333333333 cents, so we round down the result to 33 cents. Betty helped Hans investigate the impact of roundoff processes on ENIAC, and together they wrote and ran a program to see the "roundoff errors" that might be taking place. "I don't know his results," Betty admitted, "but we never used the roundoff after that."[7] Apparently, they had found a problem.

Professor Abraham Haskell Taub came from Princeton University, about an hour's drive away, eager to test the physics of shock waves. Shock waves follow explosions, and knowing more about how shock waves travel was important to military and civilian groups, including those using bombs and dynamite. Jean was happy to join his team, and so, too, was a new addition to the ENIAC programming team, Adele.

No longer teaching her numerical analysis classes and finished with the technical manuals that Grist asked her to write, Adele needed a new project. Jean was a little surprised, but happy to help a favorite teacher:

> I was assigned with her [Adele] to teach her how to program the ENIAC and to work the problem with her. Now I did not teach her how these units worked, you understand. I taught her how to do the notation, how to coordinate and get everything together to do the problem. Because I can assure you that at that time it would have taken Adele a very long time to learn how to program.[8]

Jean added, "That's why they assign somebody with another Programmer, because normally... we worked in teams,"[9] just as many programmers do today.

Jean taught Adele the pedaling sheet system she and Betty created, and together they programmed and set up Taub's program on ENIAC, debugged it, and ran it for different types of shock waves. As they came to work together, they "made a great team," Jean felt. "She was my second perfect partner."[10] Soon they also became great friends.[11]

Soon after they finished their calculations, the visitors began to speak and publish. Their problems were not confidential, and they had discoveries to share.

Hans spoke about the roundoff errors at the Moore School in the summer of 1946.[12] Douglas published his article on ENIAC in the October 1946 British journal *Nature*.[13] Although his program never worked well because of errors in the mathematics[14] (for which he apologized to Kay), he became an enthusiastic evangelist for ENIAC and devoted his article to explaining ENIAC to the publication's broad audience. Among others, he thanked "Miss K. McNulty, for information, advice and help in setting up and operating the machine on this work."[15]

They visited and wrote to each other for years and "formed a very great friendship out of the whole thing," Kay said.[16]

Taub published his results in 1947,[17] and Goff shared his results with BRL, which published them in a survey of ENIAC problems published in 1952.[18]

BRL was happy too. ENIAC was proving its value "to perform a broad range of useful work,"[19] and John Mauchly's vision of ENIAC as a general-purpose, programmable, all-electronic computer capable of solving a tremendous array of human problems was coming to fruition. Or as Marlyn said with a laugh, "this machine could do anything we wanted it to do. We were very cocky about that..."[20]

In the process, the profession of modern programming was born. A group to serve as liaisons between people with problems and computers that would help solve them. The six women were the first professional programmers of a modern computer.

The Moore School Lectures

No longer working wartime hours or preparing ENIAC for its debut, the ENIAC 6 had time to work on their hundred-year problems and catch up a bit on normal life. Jean started dating a brilliant and handsome Moore School graduate student named Bill Bartik. Like Jean, Bill worked in a restricted room on the first floor of the Moore School. In his case, the room was operated by the Office of Strategic Services (OSS), a forerunner of the Central Intelligence Agency.

Bill was an expert in "electrical noise," and in addition to working on his graduate engineering classes, he spent part of the war figuring out "how to shield delicate instruments from being affected by it."[1] A Philadelphia native, born and raised, Bill had a deep bass voice and loved to sing arias from operas and hits from musicals. Jean loved being with him, and Bill enjoyed his brilliant, outspoken girlfriend from the farms of Missouri.

Homer continued to date Fran, and they were spending a lot of time together.[2] Marlyn had met a young dentist named Philip Meltzer who was just opening up his own practice and was as outgoing as she was shy.[3] They made a good couple.

Soon it would be time to move the ENIAC south to BRL and the Proving Ground. But before that happened, the Army and the Moore

School wanted one more event while ENIAC was still in its centralized location. They wanted to host a series of lectures about the underpinnings of modern computing, ENIAC and EDVAC, with a core of technologists and scientists. The talks would start on July 8 and run through August 31.

It was a bold idea, but they needed two people no longer at the Moore School. They needed John and Pres. Despite the bad feelings, John and Pres agreed to return to the Moore School and lead the lectures. They believed deeply that modern computers were for the many, not just the few. This was an opportunity to share what they had invented, learned, and built during and just after the war. It was lousy timing for the two men, who had just opened their own computing company with the goal of making the first modern commercial computers, but it was time they were willing to dedicate.

On July 8, 1946, men arrived from some of the leading technology companies, military units, and academic institutions to learn from John and Pres. Most institutions were limited to one slot, although MIT got five and General Electric and the National Bureau of Standards got two. Bell Labs sent Claude Shannon, University of Cambridge sent Maurice Wilkes, BRL sent Samuel Lubkin. Other "students" came from the Naval Research Lab, Army Security Agency, Reeves Instrument Company and Manchester University, home to Douglas Hartree in the United Kingdom.[4]

In a full classroom on the second floor of the Moore School, John and Pres shared their insights and inventions in a series of lectures that established the foundation of modern computing. John's lectures included "Digital and Analog Computing Machines," "Conversion between Binary and Decimal Number Systems," "Code and Control II, Machine Design and Instruction Codes," and other topics. Pres delivered lectures that he titled "Preview of a Digital Computing

Machine," "Types of Circuits—General," "Reliability and Checking," and others.[5]

Some of the young ENIAC engineers got their chance to present lectures, including Chuan with "Magnetic Recording," and Kite's lecture titled "Switching and Coupling Circuits." Herman and Arthur shared a series of five sessions on "Numerical Mathematical Methods."[6]

Only those invited to attend and speak came into the crowded classroom, and that meant there was not a single woman present. Betty, who wanted to follow the discussion, sat in the classroom next to the lecture room. Because it was so hot, the Moore School Lectures classroom door was propped wide open to catch any breeze on the hot, muggy Philadelphia days.[7] Betty could hear the lectures and the discussion, even if she could not see the board or participate.

Betty would not have been allowed in if she had asked, and she would have never asked. Soon Jean pulled up a chair right next to Betty. When they were not finishing the hundred-year problems, the two women listened to the lectures that would shape the next hundred years of computing. It is too bad no one thought to invite them to speak.

The Moore School Lectures would go down in history and are still in print today.[8] John and Pres gave most of the lectures, John as the ultimate visionary and teacher and Pres as the ultimate engineer who had bypassed what others knew to be impossible. Both freely and openly sharing their ideas, knowledge, and vision rather than bottling them up or limiting access.

Maurice Wilkes would rush back to Cambridge and rapidly build his own digital computer, appropriately named EDSAC (Electronic Delay Storage Automatic Calculator) to fit with its sisters ENIAC and EDVAC.[9] He would also return to the Moore School to celebrate the birth of ENIAC in some of the anniversaries to come.

It was the best of times as men across the United States and United Kingdom gathered to share ideas that would ultimately revolutionize the world. But for some it became the worst of times. Soon after the Moore School Lectures ended, John and Mary decided to go on a quick vacation to their summer home in Wildwood, New Jersey.

On September 9, just after midnight, they decided to go swimming and walked two blocks to the ocean. They jumped in nude, feeling carefree. "It was a foolish thing to do," John would say later. "We never did anything like that before."[10]

Mary was knocked over by the waves and screamed for help. John tried to reach her twice but was pushed back by the water. He was very shortsighted and could not see much; a fog came up, further clouding Mary from his view. Naked, he ran two blocks to the first lighted home he found. It turned out to be the home of the Beach Patrol captain, who summoned the police. At 2:00 a.m. on the morning of September 10, Mary's body was found on the sand two blocks from where she had gone in.[11]

Mary was thirty-nine years old. An investigation followed, and John was questioned, but nothing came out in the interview or medical examiner's report to indicate that the death was anything but a tragic accident.[12] John was devastated. Their two children were eleven and seven.

"John walked around like a zombie for weeks, absolutely devastated by her loss." It was a horrific time: "At the time, he had no idea that he would ever be happy again . . . [I]t was a numbingly sorrowful ending to a summer full of changes."[13]

Their Own Adventures

As for the women, as they entered the summer of 1946, the hundred-year problems kept them busy, but it was also time to take vacation. Each woman accumulated vacation as Army civilian employees during the war, and although they could not take it then, the time was still waiting for them now. Jean had only a few weeks, as she joined the Army computing program late in the war, but Kay, Fran, Betty, and Marlyn had many weeks of vacation and wondered how they would spend so much time.

Marlyn headed off first to the hot spot of Havana with a fellow Computer.[1] Cuba was a place at the time that "socialites and celebrities flocked to" and ordinary Americans too. A place of nightclubs, hot jazz, and spicy food.[2] Marlyn had a very good time.

Jean left next. By August, the Taub problem was ready. Adele would finish putting it on ENIAC while Jean took Bill to Missouri to meet her family.[3]

Jean and Bill followed Jean's original path backward. They took the train from Philadelphia to St. Louis, then transferred to the Wabash Express headed west for their drop-off in Stanberry, Missouri.

For Jean it was the moment she had been waiting for. She flourished in her time as a Computer and ENIAC Programmer, and succeeded in the great city of Philadelphia. Now she had many stories to tell and a brilliant young man to introduce to her family. She could not have been happier.

Bill, for his part, was a little disoriented. Philadelphia born and raised, he had never left the East Coast and the tiny town of Alanthus Grove, farm life, and Jean's large and boisterous family were a bit of a surprise.[4]

Soon he acclimated as Jean introduced him to the members of the Jennings clan and showed him the family farm, schoolhouse, and local church. Everything was named Jennings, including the school and the cemetery.

Jean was happy. She was back with the cows and chickens, dogs and cornfields, and once again eating her mother's morning biscuits.

One night they went on a double date with Jean's youngest sister Kackie and her boyfriend. At Frog Hop, a nightclub where the big bands played, everyone was dancing except for Kackie's date, who seemed a little unhappy.

Jean asked Kackie why she was dating him and Kackie said he wanted to marry her.

"Do you want to?" Jean asked.

"No," answered Kackie.

"You should only marry someone you want to."

After that date, "Kackie wised up and dumped him," Jean recalled.[5]

The Jennings family could see the relationship between Jean and Bill was serious. One afternoon while Bill Bartik and Bill Jennings were plucking a chicken, Bill asked Jean's father for permission to marry her. Jean's father felt that was not his question to answer.

"I don't take care of my children's sparkin'," he responded. "Does she want to marry you?"

"Yes," said Bill.

"Then it's okay with me," Jean's father responded.[6]

Before they left, her father hugged Jean tight. He would be sorry to miss her wedding in Philadelphia, but as he said when she first left, his place was on the farm. She and Bill would always be welcome to come home, and he was happy they had made the long trip to be with him and the family.

Meanwhile, Betty, Kay, and John Holberton began to plan their own adventure. They had a dream of using their accumulated vacation time to drive across the country. In the age before superhighways, this meant thousands of miles of mostly two-lane roads through towns and farms, prairies and parks. After so many years in classrooms and offices, behind desktop calculators, and before big, black units of ENIAC, the three wanted an open road ahead and a broad sky above.[7]

The trip would take some planning, and they began to put their notes together and think about the routes they would take and the places they would visit.

It was time to relocate ENIAC to Aberdeen, Maryland. Built like a ship in a bottle, all thirty tons of it needed to be taken out but would not fit through the doorways and hallways of the Moore School.

ENIAC had been built in a big lab room at the back of the first floor, near the back wall. If part of the thick brick wall of this solidly built former musical instrument factory could be removed, then the eight foot tall, two foot wide, heavy steel black units could be removed, wrapped up, and taken to BRL. The women hoped the Moore School chose a good mover and that the ENIAC's units survived the bumpy roads to rural Maryland.

It would be the perfect time for Kay, Betty, and John Holberton to take their leave while the components were being moved. They piled into John Holberton's car in early October and headed west. Holberton was the driver, Betty seated shotgun as navigator, and Kay in the back seat as commentator, raconteur, and alternate driver.[8] It was going to be a glorious trip!

The Pennsylvania Turnpike, which opened in 1940, was the first long-distance controlled-access highway in the country. With John and Kay sharing driving duties, the trio traveled across Pennsylvania into Ohio, through the cornfields and apple orchards of Ohio, across Indiana, and finally to Chicago, where they visited Nick Metropolis, now back from Los Alamos and serving on the faculty at the University of Chicago. They had dinner, and a tour, and a wonderful reunion.[9]

Then on to Denver, Boulder, and Salt Lake City, staying in parks and small motels. A motel cost about $3 per night per person, and Betty and Kay shared a room.[10] A huge steak dinner cost sixty cents. The most expensive thing they did was stay for two nights at the lodge in Yosemite National Park. There they slept, toured, and "had a steak dinner," for a grand total of $11 each.[11]

Holberton's father had been an agent for the Agriculture Department, and John knew every breed of cow, pig, and horse and various crops, plants, and trees. As they passed the farms and ranches, he named the animals and plants for Kay and Betty.[12] Soon they felt pretty expert too.

Eventually, they reached the West Coast and Pacific Ocean, which none of them had seen before. They stayed near San Francisco first, spending a few nights with Kay's aunt in a suburb. After touring the area, they drove to Los Angeles to stay with another of Kay's aunts.[13]

The family owned a boatyard, and "during the war, they built PT boats [small, fast patrol torpedo boats]; now they were building gorgeous yachts made out of solid mahogany." The family hosted them royally and gave them tours of the area, including out to Catalina

Island and up to Palomar Mountain, "where we saw them polishing the mirror on the telescope there."[14]

Then farther south, they visited one of Betty's sisters in San Diego. "From there, we went down into Mexico to see some of the sights along the border," Kay recalled.[15]

In California, at a scheduled "mail drop," a location given to family and employers to send letters to be picked up when travelers arrived, they received a surprise letter from BRL. Dr. Dederick, associate director of BRL, wrote that in late October, a unit of ENIAC caught fire (someone left it running with the back panel off), and they had to rebuild it.[16] ENIAC's move was now delayed until December, and the group of three did not need to rush home.

A bit dismayed by the fire, but happy that no one was hurt, the three were not at all unhappy to have extra time on their vacation to expand their remarkable cross-country trip.

They headed back, more leisurely now, on the famous Route 66, which they found out "wasn't much of a route at that time...just gravel roads."[17] But still it took them through magnificent locations, including Arizona's Painted Desert and on to the Grand Canyon.

Next they headed southeast to the Gulf of Mexico and New Orleans. There they received another letter from Dr. Dederick, "which said that we had gotten elevated from an SP...to a P2," meaning Kay and Betty's civilian rank had been upgraded from "subprofessional" and they would start their work at the Proving Ground as "professional."[18] The two women, and John Holberton, were thrilled.

That night, they dressed up and went out on the town to celebrate! Their destination was Broussard's, opened in 1920 and still serving today, where they dined on French-Creole delicacies, including Oysters Rockefeller and "Poisson En Papillote," fresh fish cooked in parchment paper with sauce and spices. Betty remembered happily, "we went out on the town that night."[19] They finally had the rank they so richly deserved—the Army had recognized their hard work and accomplishments.

Their trip took them east and then south down the peninsula of Florida to Miami, where they visited a friend of John's and enjoyed the warm Atlantic Ocean.

Soon it was time to head home, heading north through Georgia, South Carolina, and North Carolina. In southern Virginia, they stopped to visit John's home and family farm. They stayed with his family, and he showed them around the beautiful rolling hills and farm where he grew up.

On Thanksgiving Day, Thursday, November 28, 1946, they arrived triumphantly back in Philadelphia, having covered over 11,000 miles together.[20]

ENIAC 5 in and around Aberdeen

The month of December was a busy one. There were two engagements to be celebrated, Marlyn to Philip and Fran to Homer, and one wedding. On December 14, 1946, Jean married Bill Bartik in a ceremony at a church near Betty's home, hosted by Betty's family and attended by many members of the ENIAC team.[1]

Jean planned the wedding with Betty, her best friend and now maid of honor. Betty's family hosted the ceremony, with Betty's mother preparing most of it. After all, "the Snyder family was a second home" for Jean, and now a second family.[2]

The ENIAC 6 attended, along with many men on the ENIAC team. Adele and Herman attended too. John Mauchly, still in mourning, said yes when Jean asked him to walk her down the aisle, although he teased her saying she was too valuable to be given away.[3] And no one was surprised. The ENIAC team, too, had become their wartime family.

The group took some photographs at the church. Jean in her tailored white wedding suit with long white gloves and a sheer veil reaching down to her elbows and Bill in a tweed suit with a bright striped tie. One wedding photograph shows Jean and Bill smiling happily, with John Mauchly on one side and Bill's best man, Al, on the other, and Betty in between in a beautiful dress, a little hat, and a huge smile.

Then the wedding party and guests moved from the church to the Snyder home nearby for a reception that was "fun and warm," Jean said, and "beautiful," Kay shared.[4]

And poignant. A last moment together. Everyone knew that this group, the ENIAC team, forged in the ENIAC room during and after the war, was the family they had chosen. They worked hard together and played hard together, and remained each other's companions for many months after the war. But the team members were moving on to new companies, new opportunities, to bring their knowledge of new technologies and new programming techniques to new places.

After the wedding, Betty, Kay, Fran, and Ruth prepared to move down to Aberdeen. They would start new jobs for BRL in January, following ENIAC to its new home.

"I had committed myself to go to Aberdeen. So I did," Betty recalled.[5]

Jean would continue to work for BRL on a contract through the Moore School. Dr. Richard Clippinger, head of BRL's wind tunnel projects, had a set of programs he wanted to run about supersonic flight, and he needed ENIAC to run them. On learning that an ENIAC Programmer was free, he had tracked Jean down and enlisted her.[6] She would live in Philadelphia, manage his project, and come down to Aberdeen to teach him to program. Jean was happy.

Only Marlyn would be unable to continue on BRL projects, as December 31, 1946, was her last day working with the Army. She would soon marry Philip, and he had opened a dental practice in Trenton, New Jersey, about a half hour northeast of Philadelphia. He had purchased a new building, and he and Marlyn would live on the second floor after they were married, run the dental practice together on the first floor, and rent out the third one.[7]

Had ENIAC stayed at the Moore School, Marlyn would have continued to commute to Philadelphia, but with ENIAC now farther south at the Proving Ground, the daily commute would be impossible.[8] BRL understood.

In January 1947, Kay, Betty, and Ruth moved back to the Proving Ground, this time to slightly upgraded accommodations than they had just a couple of years before. It was a women's dormitory, the rooms were still basic, and the showers were still down the hall, but someone made their beds every day and dropped off clean towels.[9] Their new professional rank had its privileges.

Betty and Kay now became roommates, with Ruth down the hall, the old friends back together. Fran moved into town. When she married Homer, they would get their own apartment.

The Proving Ground of January 1947 was physically the same as the Proving Ground they left in July 1945, but it felt so different. For one thing, they had been there in the hot, muggy summer of '45, and now the winds whipped off the Chesapeake Bay, cold and icy in '47.

They had been there when the Proving Ground was filled to capacity, 32,000 soldiers and civilians, and tens of thousands of young men drilling and marching everywhere. Now, although there were a few troops in training, most of the young men had gone home and the base was much quieter.

One thing had not changed. Their supervisor, John Holberton, was there to greet them. He had moved down from Philadelphia, too, and they would continue to work together. Kay and Betty, of course, were delighted.

Kay and Ruth were assigned to help reconstruct ENIAC. It had been delivered to the three-story brick building built to house it and other labs—APG Building 328 in the BRL section of the Proving Ground. The same building the women had studied in earlier, with its specially constructed and reinforced second floor, was now ready for the 30 tons of ENIAC, with 1,800 square feet of space, air-conditioning, and bright lights.[10] It would be a fine home for ENIAC.

But ENIAC had not arrived in good shape. Although the panel that had been destroyed by fire had been rebuilt, other units had suffered damage as they bounced down local roads from Philadelphia to Aberdeen, Maryland. It would be a tremendous job to put this computer into working order.

Kay and Ruth joined the teams. Each worked with a group of male engineers to reinstall ENIAC units one at a time, painstakingly checking first the back of the units to fix wires, circuits, and vacuum tubes. They then examined the front with its plugs, switches, and lights.

Kay and Ruth put the ENIAC through its paces with "direct programming"—they tested each unit thoroughly with the diagnostic software they had taught themselves at the Moore School.[11] They confirmed each and every operation of a unit before declaring it to be in working order.

"A Programmer that was finding the hardware problems," Kay said with a laugh.[12]

And she remembered it was slow going. "I remember many times when I was there just sitting in the ENIAC room all afternoon long, just testing decade counters..."[13]

When Kay and Ruth said a unit was ready, the teams moved on to the next unit.

Day after day, week after week, month after month, the forty unique units of ENIAC and five additional IBM units rose again in their new and permanent home.

"We did a lot of testing on it... and got it working again," Ruth said.[14]

After seven long months, at the end of July 1947, Kay and Ruth and the engineers finally emerged from their reinstallation marathon. ENIAC was up and operational. They had become good friends with the engineers but were happy to leave the ENIAC room and move on to new projects.

Betty and Fran spent the winter and spring of 1947 meeting the new Programmers that BRL had hired, a group of women and men working on problems assigned to them. In the new programming rooms, they found themselves surrounded by former Computers and old friends, including Gloria Gordon and Ester Gerston from Ruth and Marlyn's Third-Floor Computing Team, Marie Beirstein from Betty's computing team, and Lila Todd, who had mentored so many in Philadelphia.[15]

Lila Todd also brought Homé McAllister and Wink Smith because they were "two of my best employees."[16] Of the approximately 100 Computers at the Moore School, only about a dozen now remained, now promoted to professional status. The new arrivals looked forward to learning ENIAC.

But initially there was nothing new to teach them with. Like the ENIAC 6, the first Programmers at the Proving Ground had to study ENIAC's diagrams to see how its units worked: "I spent long hours trying to understand the 'blueprints' and wiring diagrams for the ENIAC and to try to learn direct programming," Homé said with frustration.[17]

While Betty and Fran did their best to explain ENIAC operations, it was still a hard way to learn. But the days of "direct programming" of ENIAC would soon over. A big change was in the wind, and Jean would be right in the center of the storm.

In the meantime, Fran married Homer in March 1947. After their honeymoon, both spouses continued to work on ENIAC, Homer as its trusted maintenance engineer and Fran as a valued Programmer.

In February, Betty began spending all her weekends in Philadelphia. Not for vacation or fun. John Mauchly and Pres had a new company and were designing a new machine. It sounded exciting to Betty, but the young company had little money. Still Betty spent the weekends there, "for nothing," just to help the most exciting project she knew.[18]

On Friday and Saturday nights, she stayed with her parents. She worked with John and Pres on Saturday staying until late in the night, and she worked Sunday until the last train to Aberdeen. By the time each Monday morning rolled around, Betty would be back at BRL.

Betty worked closely with John and Pres. They brainstormed about the new computer and "what we thought the machine should have in it."[19] The talks were exciting and consuming. The new computer would soon be named UNIVAC, the world's first modern commercial computer. So many UNIVACs would be delivered across the United States that "UNIVAC" would become synonymous with "computer" for years to come. But not yet. The first UNIVAC had not yet been built in 1947. It was still a concept, and one for which the banks were reluctant to provide funding.

Meanwhile, Richard "Dick" Clippinger, head of the BRL's wind tunnel project and a mathematics professor at Harvard, wanted Jean's help with his wind tunnel equations.[20] He wanted to use ENIAC to simulate the performance of a plane at supersonic speeds. Planes had not yet broken the sound barrier, and there was a question whether the planes and their pilots would survive.

Dick was a brilliant mathematician and expert on wind tunnel equations, but he did not know anything about programming ENIAC. Jean did, and Dick hired her to direct his programming group and program his equations. The contracts went through the Moore School, and Irven Travis hemmed and hawed until Jean told him she had no intention of getting pregnant, at which point he quickly signed the papers.[21]

Jean was the manager of the project. Her first order of business was advertising for, interviewing, and hiring four new Programmers. Since it was a new process, and no one had advertised for a modern programmer, she made it up as she went along. She chose the candidates who "interviewed me about the job rather than my interviewing them about their qualifications." She chose those with "energy, curiosity, and sense of adventure."[22]

She hired two women and two men, Kathe Jacoby, Sally Spears, Art Gehring and Ed Schlain.[23] Each member of her team continued on to lead long and successful careers in computing, but first Jean had to teach them about ENIAC and programming.

At the same time, Jean was commuting to Aberdeen to teach Dick how to program ENIAC. In turn, he was teaching her about wind tunnels, large tubes with air moving inside used to learn more about how objects moved, including airplanes and rockets. Scientists often used scale models in wind tunnels, and sometimes full-size models.[24] ENIAC would simulate the wind tunnel testing process.

Soon, however, Jean and Dick realized they had a problem. ENIAC was too small for Dick's wind tunnel equations. Even with all the tricks Jean knew, and all of the ways she, Kay, and Betty had figured out to optimize the trajectory program, the wind tunnel with its "hyperbolic partial differential calculus equations" was simply too big to fit on ENIAC.[25] The computer did not have enough room for all the program pulse cables, digit wires, and switches that the program would need.

Instead of giving up, they decided to change ENIAC. And the way to do it was planted by John von Neumann, who had brought the Los Alamos problem to ENIAC.

John von Neumann suggested to Dick that "it would be possible to run the ENIAC in a way very different from the way contemplated when it was designed."[26] They could repurpose ENIAC's portable function tables to store instructions, rather than numbers. Then those instructions could be fed into a specially organized ENIAC, and space would no longer be a problem.

Thus began one of the busiest periods of Jean's life. In the spring of 1947, she and Dick began commuting up to Princeton's Institute for Advanced Study to meet with John von Neumann and Adele. Their mission was to work together to convert ENIAC into a "stored-program" computer, albeit an unusual one.[27] Unlike EDVAC being built at the Moore School and EDSAC being built at Cambridge University, ENIAC would not be built from the ground up as a stored-program computer, but would be reengineered to become one. That's a little like doing backflips and cartwheels at the same time.

In the mornings, Jean, Dick, and Adele worked with John von Neumann on the first floor of the Institute to figure out what instructions would be needed in an optimal instruction set.[28] Albert Einstein had his office there too.

In the afternoon, Jean and Adele would go off to sketch out the details of the instructions they had discussed that morning. The two roughed out how these instructions might be put on the ENIAC by direct programming. "We were working on implementing the code" and figuring out roughly how to program these new code instructions onto ENIAC, Jean reflected later.[29] The pair were happy to be working together once again.

In the afternoons, Jean and Adele sometimes worked in an open room of the Institute or in one of the odd Quonset huts that the Army had constructed on the Institute's grounds during the war.[30] The now-falling-down steel structures existed much to the dismay of the neighbors, who thought they were unsightly, ramshackle eyesores in their beautiful neighborhood.[31]

Herman and Adele had set up residence in one Quonset hut. Jean and Dick would sleep in two others when their work stretched late into the night, or over a few days, as it often did.[32]

After these meetings, Jean returned to Philadelphia, and with her four-person team, she worked out the specific details of the instruction codes.[33] They figured out exactly how these would be put on ENIAC, and what units, switches, digit cables, and trays and program pulse cables would be used.

If Jean and her team figured out that an instruction did not work or was too large and took up too many resources on ENIAC, leaving insufficient room for other instructions, then Jean returned to Princeton to rework or rewrite it.[34] Like Kay and Ruth in the reinstallation of ENIAC, Jean was trusted for her evaluation and testing. Until she approved, the instruction was not accepted.

When Jean returned to Princeton with a problem, John von Neumann, Adele, and Dick would listen carefully, and often John "would propose a simpler version of the instruction or a substitution of the instruction."[35]

It was a meeting of equals, except when it wasn't. On one occasion, the team was talking and when John von Neumann proposed a new instruction, Jean said no, as she knew this particular idea would not work. Herman, who was always in the morning discussion but "never made any contributions," Jean recalled, "glared at me as if I had blasphemed God."[36]

But John von Neumann "merely laughed" because he realized his mistake and "proceeded to correct himself."[37] The group continued forward with their work.

By the middle of summer 1947, Kay and Ruth and the engineers were finishing the installation of ENIAC, and Jean and the Philadelphia-Princeton-Aberdeen team were finishing ENIAC's new instruction code, named the "Converter Code." Jean and her team delivered the programming sheets for the Converter Code to Dick Clippinger and BRL on paper.[38] They did not travel down to Aberdeen to set up ENIAC to run program—someone else would use their work to transform ENIAC into the "full-fledged stored program computer" that it would soon become.

Kay and Ruth went off to find new programming tasks, and Jean and her illustrious team finally turned their attention to the wind tunnel equations that Dick was waiting so patiently for.

A New Life

In August 1947, Kay emerged from the ENIAC room. She hoped to begin work on another program for Douglas Hartree. He had reworked the mathematics of his calculations and wanted Kay to reprogram it.[1] She looked forward to doing so, but found another problem and opportunity more immediately in front of her.

What she found was a group of new Programmers milling around without much direction, and then a copy of Adele's manual lying on a desk at BRL. The manual was a fine discovery and laid out each unit of ENIAC, how it worked, and what it could be used for.

"It was very good, but long, long overdue," Kay noted.[2]

She gathered the Programmers in her area around her and started teaching them, from Adele's manual and her own experiences.

Jean's team went on to their next set of deliverables, using the Converter Code to program Dick's wind tunnel equations. They worked closely with two young theoretical mathematicians, A.S. Galbraith and John Giese.[3]

Following the pattern that Jean and Betty had used on the trajectory, where Jean did the mathematics and Betty did the logic, Galbraith and Giese broke down the complex equations into smaller pieces, and Jean's team broke the equations down into the individual operations ENIAC could perform.[4]

For the latter half of 1947, Jean traveled regularly to Aberdeen to work with Dick, Galbraith, and Giese. Dick, in turn, came up regularly to Philadelphia to cheer on Jean's team and watch as his wind tunnel equations came to life.[5]

Continuing ENIAC team traditions, Jean, Dick, Kathe, Sally, Art, and Ed ate lunch at Lido's Restaurant on Woodland Avenue and shared their excitement about computing. Dick regaled them with stories of his travels and living abroad. He had ridden the Trans-Siberian Railway across Russia, studied at the Sorbonne in Paris, and traveled around Europe.[6]

Jean was happy. "He always complemented my team and we all adored him," she wrote. For her part, Jean was having "so much fun working with the ENIAC, I thought I'd died and gone to heaven" and "I couldn't wait to go to work every day, and it was just wonderful."[7]

But all good things must come to an end. Eventually, Dick's promised wind tunnel equations were done, neatly programmed on their programming sheets, too, and timely delivered to Dick. "We met every goal," Jean proudly wrote.[8]

Jean's team knew they had done something important for Dick and for BRL. "By completing the project, my team had given birth to a new mode of programming that—like the ENIAC itself—would forever change the world."[9]

Hans Neukom, in his 2006 article for the *IEEE Annals of the History of Computing*, called it "a second life for ENIAC."[10]

Betty quickly became expert at programming in its new mode. In the summer of 1947, she was asked to write "a suite of test programs" for ENIAC using the new Converter Code and did so.[11]

Calm, cool, and meticulous, Betty was the right person to create these new test programs, and for the rest of her life, people would come to her to test new codes and programming languages, including those that came later, such as the powerful Fortran and COBOL.

Kay trained the new Programmers around her and then worked on Hartree's problems.[12]

Ruth became a senior Programmer and leader of many programming teams.[13]

Homé McAllister was placed on the team to prepare and run additional wind tunnel equations for Dick Clippinger.[14]

Dick, in turn, published the first paper about the Converter Code for BRL in September 1948. He credits John von Neumann, Adele, Jean, and himself for the "finished regime" of the Converter Code. It notes contributions of Galbraith, Giese, and Jean's team of Kathe, Sally, Ed, and Art, among others.[15] They were happy to be acknowledged for their hard work.

The press continued to beat a path to ENIAC's now-open door at the Proving Ground and sent their best photographers to Aberdeen to capture man and machine, often woman and machine. *LIFE* magazine's photographer Francis Miller captured Betty with her foot confidently resting on a lower bar of a function table, clearly master of the great computer.

Another beautiful image captured Gloria Gordon and Ester Gerston in front of ENIAC. Ester stands with black digit wires draped over her arm, and Gloria crouches as if to plug in a program pulse cable at the bottom. Republished in the *Washington Post* when she passed away in 2009 at age eighty-seven, the caption includes the women's names and Gloria's snappy remark about the world-famous image: "Great! Now everyone gets to see my big butt!"[16]

In many photographs, Ruth is present and unnamed as she stands quietly in the background, now supervisor watching younger Programmers at work. In 2019, Ruth made the front cover of the *New York Times Magazine*, but still without her name.[17]

Soon it was time for the ENIAC 5 to leave their jobs with and supporting BRL. They had done their job well, helped put ENIAC back on its feet, trained new groups of Programmers and mathematicians to prepare equations and programs for ENIAC, and created and taught the new Converter Code to take ENIAC to the next level.

Fran was pregnant, and she chose to leave her job with the Army to focus on motherhood.[18]

And John and Pres had gotten their first customer and an advance payment. They could now afford to hire Betty, and she joined them in June 1947 as employee number thirteen of Electronic Control Company, soon to be renamed Eckert-Mauchly Computer Corporation.[19]

By March 1948, Jean's whirlwind year at Princeton, Philly, and Aberdeen was over. "The contract was up and we had completed the twelve goals that it called for," Jean wrote.[20] When Dick wanted to enlist Jean and her team for another contract, she passed the programming pen to Art, who "did it very well." Jean was ready to move to her next project. She was ready for a new adventure.

Ruth, bored on the weekends in Aberdeen, traveled back to the Rebecca Gratz Club, where she had lived, to attend weekend parties. It was there she met Adolph Teitelbaum. They hit it off and she later found out that he was good friends from college with Marlyn's husband, Philip.[21] The four spent many happy times together, double-dating to concerts, plays, and dinners. Soon Ruth and Adolph were engaged.

They married in 1948 and took a long honeymoon traveling

cross-country by car through the Southwest. When they returned, Ruth applied to the Moore School to work in Philadelphia and Irven Travis had no hesitation. She stayed for a year and a half, but the couple had fallen in love with the Southwest and its clarion call beckoned them. They moved to Dallas, Texas, and stayed for the rest of their lives.

Kay, for her part, intended to remain in Aberdeen for a while. She was happy, challenged, and had many friends. She enjoyed programming, teaching new Programmers, and using the Converter Code.

Plus, she was working on a special project for John Mauchly. Somehow, this overworked president of a young tech start-up got stuck making the transcripts of the Moore School Lectures. Recorded on early magnetic recording wire, he was supposed to listen and write down every word. He was the single father of two small children and had a company to run.

When he shared the problem with Kay, she urged him to bring the recorder and magnetic wire to the Proving Ground and she would help him with the transcripts.[22]

From then on, John stopped by the Proving Ground when he was en route to Washington, DC, to court government customers for UNIVAC. Initially he stopped to check on his project, and then he stopped to check on Kay. In late 1947, John asked Kay to marry him, and Kay said yes. Now she, too, would move back to Philadelphia.

Sorry to see her go, in February 1948, her old teammates, the men she worked with rebuilding ENIAC, threw Kay a bachelorette party. The next day, they carried her suitcases to the train platform and hung signs on them—ABOUT TO BE MARRIED—for all to see.[23]

Kay and John were married in a small church ceremony in

February 1948. All of the ENIAC Programmers were in attendance, and Pres was the best man.[24]

ENIAC remained in active use at the Proving Ground until October 1955. In its new form using the Converter Code, and upgraded from time to time, ENIAC ran over 100 important problems.[25] In his survey of ENIAC's operations and problems from 1946 to 1952," second-generation ENIAC Programmer W. Barkley Fritz wrote, "the [Converter] code has improved the production operation of the ENIAC in the following ways," and laid out that it made it easier to prepare problems, allowed larger programs, reduced the time for changing programs, and simplified testing procedures.[26]

The Converter Code made ENIAC easier to use, and that opened the door to larger and more complicated problems programmed by a wider group of people. The world beat a path to BRL's doorstep, and BRL graciously allowed in universities, government agencies, and other military groups to run a wide range of problems.[27]

We know only the nonclassified ones, and they include weather forecasting, thermal ignition (burning of powder for weapons), rocket trajectories, mathematical models for well depletions (US Bureau of Mines), and the calculation of a special wind tunnel design for MIT.[28]

Even scientists and mathematicians from Los Alamos returned for more calculations, with Nick Metropolis, Klara and John von Neumann, and Adele all helping.[29]

For several years, ENIAC would be the world's only supercomputer, "if the reader accepts the definition of a supercomputer as the most powerful and expensive computer available," Barkley Fritz wrote.[30]

The teams used the Converter Code that John von Neumann,

Dick, Adele, and Jean had so carefully designed and that Jean's team had so carefully programmed. There would, of course, be tweaks and expansions over time, but the core framework stayed the same.[31]

The ENIAC's future Programmers gained much from the Converter Code, but lost something too. Gone were the swashbuckling days when ENIAC's original Programmers controlled every microsecond of the program, created powerful parallel processes, and engaged in the mind-boggling job of tracking every unit's individual timing to make sure each result reached the next unit at the right moment. The ENIAC 6 were among the lucky few to ever use ENIAC in this way.

In its Converter Code adulthood, ENIAC became four times slower, as it took a long time for instruction codes to travel from function tables to the main units for processing. And it became "serial," meaning it was able to process only one instruction at a time.

Only a handful of people ever used the ENIAC in its original, powerful, "direct programming" mode, and only six were ever employed solely for this task:

Betty Holberton, née Snyder
Jean Bartik, née Jennings
Kathleen Mauchly, née McNulty
Ruth Teitelbaum, née Lichterman
Marlyn Meltzer, née Wescoff
Frances Spence, née Bilas

Only the ENIAC 6.

Epilogue

The women in the black and white photographs of ENIAC stayed with me. When I returned to the Harvard campus from the Computer Museum in my junior year, I dove into research, determined to discover the truth about the women in the photographs. I wrote my class paper on the mystery I found: both the history of software and the history of women seemed to be missing from computer history. I have since forgotten the title, but my subtitle was "Men Are Hard and Women Are Soft."

I took a year off before my senior year. I was a little overwhelmed at finding such a big missing chapter of computer history. How could historians have missed it? Before I wrote my senior thesis, I needed some time to think.

In the middle of the year, I found a book by Joel Shurkin titled *Engines of the Mind*, a history of computers that featured information from his interviews with John Mauchly and Pres Eckert when he was a reporter with the *Philadelphia Inquirer.* In one chapter about ENIAC, he wrote about challenges facing the ENIAC team. Then Joel dropped a footnote that startled me:

> For reasons no one has been able to fathom, the best early programmers in computer history, going back to the countess of

> Lovelace, were women. This is true not only at EMCC but in almost every computer lab. [Grace] Hopper and [Betty] Holberton are credited with writing one of the assembly languages, a basic form of communicating with computers, a significant achievement.[1]

He shared my conclusion!

I started my senior year at Harvard with a new determination and began looking for a professor to advise my senior thesis on the ENIAC Programmers. I had a hard time because I was a social studies major, and what did social theory have to do with computer science?

Fortunately, I was taking a computers and ethics course with Professor Joseph Weizenbaum of MIT. (Harvard students can cross-register, and it's only two "T" stops from Harvard Square to MIT.) The course was fascinating, and I started going to Professor Weizenbaum's office hours in a building at the far end of campus that also housed the young, and then little known, MIT Media Lab. One day I mentioned to Professor Weizenbaum that no one at Harvard would advise my thesis on the women of ENIAC. To my surprise, he looked up from his big wooden desk, smiling beneath his thick gray mustache, and said, "I'll advise it." I warned that my search might prove fruitless. Once again he smiled. "We'll see what you find," he said.

Thus began my search for the ENIAC Programmers. As I found bits and pieces of the story, I went back to Professor Weizenbaum's office and shared them. When I could not find a good description of the differential analyzer, he took out his pad and pencil and sketched what he called the "giant's mattress spring."

My secondary sources, such as the *Encyclopedia of Computer Science*, offered nothing, so I dove into primary ones. Herman Goldstine's autobiography, *The Computer from Pascal to von Neumann*, proved useful.

On page 202, he wrote about hiring six Computers to program ENIAC and named them (with a few misspellings in their names). I read page 202 again and again, and every footnote and bibliography reference, but found no further references to the ENIAC Programmers. But at least I had their names.

In early fall 1986, I decided to call the Moore School. I bounced around from one dean to another, from one professor to another. None knew anything about women working on ENIAC, until someone connected me to the office of Saul Gorn, a retired senior professor with an office still at the school. He listened carefully as I shared the pieces I had put together that indicated there were women working on ENIAC as Programmers during WWII.

He paused for a long time before he responded, slowly telling me that during WWII, he was a student at the Moore School. Although women were not allowed to be faculty or students, he remembered that there were young professional women working on wartime projects.

"They may be the ones you are looking for," he said slowly, and informed me of the fortieth anniversary of ENIAC that would take place that October. If I wanted an invitation, he would get me one. I did and he did.

With a small grant from Radcliffe College, I jumped on the Amtrak train from Boston to Philadelphia and arrived at the ENIAC's fortieth anniversary party. I found my way from the train station to the Towne Building and walked up its front stairs. After I picked up my badge, I went inside the large rectangular room that was filling up quickly. I sat down in one of the chairs lined up in front of a small stage.

The proceedings were formal. The event seemed dedicated to giving awards to various men on the ENIAC team, and initially men, young and older, spoke about the role of ENIAC and its team in the birth of modern computing.

Then Kay took the stage and leaned forward. Her speech, "Reminiscences," was short and sweet and captured the attention of the audience. She talked about her days working on ENIAC and how much she enjoyed working with people in the room for the event, women and men. She spoke with a smile and with grace; everyone listened closely, and many of the older attendees nodded at parts of her stories. It was clear she had been a member of the ENIAC team.

After the ceremony, attendees mingled and chatted, mostly men in the front. So I walked the long length of the room and spotted a cluster of a few older women in the back. They had white hair and pearls around their necks, and they were laughing and engaged in a vigorous conversation. I asked if I might listen and they said yes and resumed their discussion.

The women were talking excitedly about something technical. I did not understand the words they were using—*accumulator, program pulse, master programmer*—but I could tell they were telling a story about something in which they were deeply and personally engaged, a technology story that happened some time ago.

I learned later they were discussing the last bugs in their ballistics trajectory program before Demonstration Day, and how they had fixed them. They shared their memories. When they debated a few details back and forth, it seemed very collegial. These women clearly had a warm friendship and deep respect for each other. They enjoyed telling this forty-year-old technical story, and their roles in it. I was intrigued.

After they finished, they introduced themselves. They were Betty, Jean, and Marlyn—three of the women mentioned in Herman's book. Between Kay in the front of the room and the three in the back, I had just met four of the original six ENIAC Programmers named by Herman Goldstine.

I returned to Boston with new phone numbers and, over the next few months, interviewed the four women by phone, as well as Jean Sammet, who worked with Betty on the committee that designed the first cross-platform programming language, COBOL. My senior thesis emerged, titled "Invisible Women: A Social History of Contributions of Women to Early Computer Programming."

As I finished my senior thesis, I knew I had touched only the tip of the iceberg of this story. When I shared my concern with Professor Weizenbaum, he smiled and said, "Opportunity never knocks once." He knew there would be more time to research and tell this story.

After graduation, I started my own career in the highest-paying job I could find, in information technology (IT) on Wall Street. I joined Morgan Stanley's Information Services Management Training Program and started to learn about the computing and programming that helped run one of the world's largest financial companies.

We studied about four hours a day and then spent eight hours a day working an IT assignment. Mine was helping run the international data networks of Morgan Stanley that connected the huge New York databases at my location to offices in London, Tokyo, and, soon, Zurich and Hong Kong. When someone in Tokyo or London had no response, they called my group and we figured out where the problem was—sometimes a satellite uplink on Long Island wasn't working, or a "local loop" of the telephone company in New York or Tokyo was down. Every day was like a puzzle, and I enjoyed solving them.

But some of my peers were less happy, particularly women. They worked in other groups that did not treat them as well, or with young men who did not think young women could make good computer professionals. The women were very good at their jobs but feared they did not belong in computing. They were tired of being told that computing was a field for men.

During our 8:00 p.m. to 8:00 a.m. shifts, when we took a break in the middle of the night, we would sit between long, high, narrow, metal racks of the huge data center while I told stories of the ENIAC Programmers. I shared with my peers that we were not the first women in computing. Rather, we followed a long and distinguished line of women dating back to ENIAC. I shared the names and stories of the programming pioneers I knew—Betty, Jean, Kay, Marlyn, Ruth, and Fran. Too soon our break would be over, and we would continue again the next night.

The stories sank in and I watched as my friends returned to their positions in the data center with their heads a little higher and their smiles a little brighter.

I briefly became a data security auditor for Price Waterhouse and then, wondering what laws would regulate all the data we were shipping around the world, went to law school. I loved law school, after I got the hang of it, as it was very different from programming. I joined the telecommunications law firm Fletcher, Heald & Hildreth and specialized in the law of microwaves, satellites, television, and radio. I began to work on early Internet law and policy problems.

But two years after I started work at the firm, in the fall of 1995, a clock went off in my head. I realized that the *fiftieth anniversary* of ENIAC was coming up in February. Would it be celebrated in 1996? I picked up the phone once again and called the Moore School. After only two calls, I reached Dean Steve Brown, who was responsible for the upcoming anniversary and who was glad I called. The ENIAC anniversary, he announced, would be celebrated in grand style in February 1996, and he was hoping Vice President Al Gore would join the ceremony. Hundreds were being invited to the big event, which would include a few days of lectures and a large dinner.

I told Dean Brown that I had written my undergraduate thesis on the six women who programmed ENIAC. "Are they still alive?" I asked. "Are they coming to the anniversary?"

His answer changed my life: "Who . . . who are you talking about?" Except for the women who were widows of the inventors or engineers, the ENIAC Programmers' names had been forgotten. They were not even on the invitation list to the fiftieth anniversary. The story of the ENIAC Programmers was about to be lost.

I asked my law firm for a change of schedule. I would work four days a week and spend the fifth day trying to find the ENIAC Programmers.

On the eve of the fiftieth anniversary celebration, I drove through a snowstorm from Washington, DC, to Philadelphia with Betty Holberton and her friend Donna Duvall in the car. Betty had suffered a stroke a few years earlier and was initially reluctant to travel, but Donna and I persuaded her to attend. When she arrived, Betty was welcomed warmly by many members of the computing community.

Because only Kay and Betty were invited to the VIP reception, I hosted a VIP event of my own. In a room of the hotel, we held our own reception before the banquet, one for all of the ENIAC Programmers, their guests, and photographer Steven Falk. Everyone posed for pictures: Betty and John Holberton, Kay, Jean and her daughter Jane, and Marlyn and Philip Meltzer. Then the ENIAC Programmers took pictures together, with their family members, and with me.

We had another visitor to our small reception, Tom Petzinger of the *Wall Street Journal*. Like me, he had called the Moore School asking questions about stories he had heard of women working on ENIAC. Because Penn had no answers, Kathy Wohlschlaeger, a wonderful organizer of the fiftieth anniversary, pointed Tom to me. I invited him to our reception and shared that I expected the women to dive into stories of programming ENIAC fifty years earlier.

They did and Tom whipped out his notebook and began taking notes. Nine months and a lot of discussions and interviews later, including with the ENIAC Programmers and me, Tom published two columns in the *Wall Street Journal* about the ENIAC Programmers. He used his weekly column about Silicon Valley technology entrepreneurs to share the stories of the ENIAC pioneers whom he thought current entrepreneurs should know.

Following the fiftieth anniversary, life went quickly. Mitch Kapor and the Kapor Family Foundation gave me a grant to research and then film extensive oral histories of the ENIAC Programmers. I spent six months in the Library of Congress, with my own desk and shelf to house the books about 1940s and '50s computing and programming that the libraries extracted from the bowels of the library and brought to the Science and Technology Reading Room.

In 1997, I appeared with David Roland of the Roland Company; Sheila Smith, director of photography; and Mary Keigler, sound engineer, at the homes of Kay, Marlyn, and Jean, near Philadelphia, and at the library of Betty's senior living center near Washington, DC. David directed the shoots and I conducted the interviews. I came with many pages of questions neatly typed out on long legal sheets. These interviews became the basis of my 2014 documentary, *The Computers: The Remarkable Story of the*

ENIAC Programmers, and this book. Funding for the documentary came through Megan Smith and from the foundations of Lucy Southworth Page and Anne Wojcicki, and I am forever grateful.

Gradually news of the ENIAC Programmers began to spread. Tim Bartik, Jean's son, nominated them in 1997 for induction into the Women in Technology International Hall of Fame; Dr. Barbara Simons, president of the Association for Computing Machinery, named Kay and Jean as keynote speakers for the Turing Awards in San Francisco in 1999; and the Computer History Museum (CHM) named Jean a "fellow" alongside Bob Metcalfe and Linus Torvalds in 2008. (I nominated Jean but did not tell her until after the chair of CHM called to invite her to the ceremony.) Jean invited me to join her at the black-tie ceremony in Silicon Valley, and CHM graciously hosted five generations of her family.

The next night, CHM hosted a VIP reception for Jean and then a big event, a "fireside chat," with Linda O'Bryan, founder of National Public Radio's *Nightly Business Report*.

But would anyone come? The story of the ENIAC Programmers was not well known. Brad Templeton, my friend and longtime chair of the Electronic Frontier Foundation, and I ran our own publicity campaign, informally titled "Bring Your Daughter to Meet Jean." Brad introduced me to female bloggers of Silicon Valley, and they spread the word.

That night we held our breath, still not knowing if anyone would show up. Instead we found ourselves in the museum's biggest room, set up with 400 chairs, full of women programmers and engineers, some with their daughters, and some men too. Our hosts beamed—it was the youngest and most female audience the CHM had ever hosted.

We showed my trailer of the documentary (as a work in progress), and Linda O'Bryan began to interview Jean. They both sat in large, wooden, upholstered chairs on a small stage set up front. At one point, all of the lights in the room went out. When I checked later, the tech crew told me that the lights were on motion detectors. They had gone out because 400 people, hanging on Jean's words, had stopped moving to listen closely. The crew members smiled. That had never happened before.

When Kay came as a guest to the Grace Hopper Celebration of Women in Computing, Chicago 2004, she was a celebrity, too, as young women looking to make their careers in computing programming tried to learn more about her and her career.

But obstacles remain. When I tried to explain my research to the computer history community, I got a harsh response. In conversation, Dr. William Aspray accused me of "revisionist history" and changed the topic. Even still, I watched with pride as some older historians rushed to interview the ENIAC Programmers late in their lives. Michael Williams's face lit up in 2009 in Savannah when he interviewed Jean, winner of the IEEE Computer Pioneers Award.

More recently, this sexist opposition has been taken up by young computer historians who seem to feel they have to set the record straight and stamp out stories of early programming. In 2010, Nathan Ensmenger published the book *The Computer Boys Take Over.* Its cover depicts a white man standing in front of a mainframe computer. He demeans the ENIAC Programmers as "glorified clerical workers" and says without citation that the "coders were obviously low on the intellectual and professional status hierarchy."[2] Would that Kay, Jean, Betty, Marlyn, Ruth, and Fran were here to refute these odd and unfounded contentions.

And Thomas Haigh and his coauthors who wrote *ENIAC in Action* actively dismiss the value of work of the ENIAC Programmers. They refuse to bestow on them any title higher than "operator," deny the innovation and depth of their mathematical and computing work, and refuse to acknowledge the value of their oral histories. They are part of a group who actively demean those who try to tell the ENIAC Programmers' stories, including Walter Isaacson and me.

Cory Doctorow, science-fiction writer and digital rights leader, laid out his concerns about Nathan's and Thomas's work in a Boing Boing article in 2019. He wrote that computer historians young enough to be the "Programmers' grandchildren" have taken it upon themselves to write "a sexist, revisionist history of the early computer science," one in which "women invented modern programming and were then written out of the history books."[3]

Telling this missing history certainly would be a lot easier without such discriminatory pushback. How can oral histories be bad? I watched and learned from Ruth Edmonds Hill, a pioneer in oral histories at Radcliffe College, as she conducted and then transcribed and shared the interviews of the Black Women Oral History Project.[4] She interviewed elderly African American women whose experience had not been recorded or even deemed worthy of historical importance.

We watched her capture immensely important missing histories and fill in major gaps. Ruth Hill taught us that writing history requires us to collect and engage in oral histories if we want to capture the stories of women and minorities that historians so often ignore. Quite commonly, those from underrepresented groups do not leave written records. It is up to oral historians to fill in the gaps and share the important stories and lives left out.

I wish MIT Press would look at its books more closely and resolve not to use titles like *Men, Machines, and Modern Times* (1968), *A Few Good Men from UNIVAC* (with photographs showing very clearly that it was a few good women and men) (1990), and *The Computer Boys Take Over.* These titles turn away from the very young women who computer science and computing professions are trying so hard to attract, as well as those men who are not seeking such an exclusionary environment.

Fortunately, John Mauchly and Pres Eckert did not feel or act in a sexist manner. They hired, trained, taught, listened to, and encouraged the invention of anyone who could think "outside the box" and was willing to work long hours. This included women and men, immigrants and people of many religions and races. They wanted the best and the brightest and found them regardless of race, religion, or gender. My friends in senior positions of companies across Silicon Valley tell me that this type of inclusiveness and diversity is key to the success of the best tech projects today.

The ENIAC Programmers are my role models, and they were my friends. At a time when I wondered whether, as a young woman, I should stay in my computer science classes, they gave me inspiration to do so. I credit my career in Internet law and policy and as a founding member of the Internet Corporation for Assigned Names and Numbers (ICANN) to the encouragement I found in their work. I know from the tearful reactions of young women from Google, Microsoft, and Amazon to my documentary premier at the Seattle International Film Festival that the ENIAC Programmers' story inspires them too. I hope that the story we tell here in this book gives everyone permission to enter the fields of computing and programming or to seek any career of their choosing. During WWII, we learned that the country needed to use its best talent for important work. We still do today.

Postscript

I was fortunate to know four of the ENIAC 6 quite well. They shared their oral histories, insights and stories, and homes with me. They became my mentors and friends, and they met and inspired my children. I will never forget a visit to Marlyn's house where my two children, quite young, banged happily on her baby grand piano, a bit to my dismay and to laughter from Marlyn.

Each of the ENIAC 6 lived very full and rich lives. They raised families, and many continued their careers and community involvement. Below is a sample of their activities after their ENIAC work—just a sample.

Kathleen ("Kay") McNulty Mauchly Antonelli (1921–2006)

Kay worked closely with John Mauchly after their marriage and talked with him every night after work about his inventions, challenges, and goals for the Eckert-Mauchly Computer Corporation and later companies. Although she did not formally reenter the workforce, she edited most of his writings and spent a lot of time with him talking about programming and programmers. "He really believed that women… made much better programmers than men." Kay said later, "I think maybe he was right about that."[1]

In February 1950, they found a big eighteenth-century farmhouse in Ambler, Pennsylvania, a once-rural area outside of Philadelphia that is now part of its suburban sprawl. It was surrounded by fifty acres of land, and they named it Little Linden Farm. There they raised a large family, their five children Sally, Kathy, Bill, Gini, and Eva, and John's two children, Sidney and Jimmy. "It was absolute heaven," Kay said.[2]

There was plenty of room for visitors, and the computer world beat a path to Kay and John's doorstep and dining room table. Douglas Hartree and his wife spent time whenever they visited from the United Kingdom, as did Maurice Wilkes, who had built his own stored-program computer in the United Kingdom after attending the Moore School Lectures. Nick Metropolis and Stan Frankel dropped by whenever they were in the area. Even Navy Captain Grace Hopper joined the Mauchly clan for dinner.[3] Kay was an active parent, leading Girl Scout troops and a Cub Scout den while substitute teaching occasionally at her children's schools.

Honeywell, Inc. v. Sperry Rand Corp., a lawsuit to invalidate the ENIAC patent filed in 1967 and decided in 1973, took a toll on Kay and on John. This was a trial far from home, in Minneapolis, and John was very ill while it was taking place. Also, Kay was upset because she felt Sperry Rand never mounted a good defense of John and Pres and their work. The story of the trial is told in the 1999 book *ENIAC: The Triumphs and Tragedies of the World's First Computer*, by former *Wall Street Journal* reporter Scott McCartney.

In the mid- and late 1970s, Kay joined John in meeting the young hobbyists building PCs. It was the first time she began to tell her own story and her work in early programming. After John's passing in 1980, she continued speaking and talking about her work and the legacy of ENIAC. Her short speech, "Reminiscences," at the fortieth anniversary of ENIAC in 1986 described how odd it was to be a woman at the Moore School in the early 1940s and how much fun she had on the ENIAC team of women and men in 1946.

In the mid-1980s, Kay became engaged to Severo Antonelli, a famous photographer, and decided to invite her friends, the ENIAC Programmers, to a small reunion. It was forty years after Demonstration Day, and they all came: Jean from Philadelphia; Marlyn and Philip Meltzer from Yardley, Pennsylvania; Betty and John Holberton from Potomac, Maryland; Fran and Homer Spence from Syosset, New York; and Ruth and Adolph Teitelbaum from Dallas, Texas.

In 1996, Kay and her fellow ENIAC Programmers were inducted into the Hall of Fame of Women in Technology International (WITI). Although she could not attend, the event and its publicity led to many invitations and Kay began to speak around the country, often with Jean. They visited and talked at the Microsoft campus in Seattle, WITI events in Boston and New York City, and many Philadelphia events.

In the early 2000s, an Irish documentarian discovered Kay and invited her to come to Ireland to film a documentary. Kay enthusiastically agreed and invited Jean to join her. They enjoyed the trip immensely, and Kay spoke at the University of Limerick, "where our photographer and a crew met us and videotaped us"; Dublin University; and the Letterkenny Institute of Technology. "I had a wonderful time," Kay wrote. "They also named an award for me. The award goes to the best student in computer science each year."[4] She could not have been more pleased.

Kay died in 2006, and her daughter Gini now gives speeches at the honors her mother continues to receive. In 2017, Gini traveled to

Dublin City University for the renaming of its computer science building in her mother's honor, now the McNulty Building, and in 2019, she went to Chicago as Kay was inducted into the Irish American Hall of Fame.[5] Gini is Senior Director of Research and Data Management, Institutional Advancement at Chestnut Hill College, where her mother once attended.

Ruth Lichterman Teitelbaum (1924–1986)

Ruth returned to the Moore School after her work in Aberdeen Proving Ground in late 1948, but she stayed for only a short time. During their honeymoon early that year, she and Adolph had toured the Midwest and the Southwest, and fell in love with Texas. When the opportunity arose for Adolph to start a business in Dallas, they both jumped "at the chance to run our own business" and move to a new part of the country.

Adolph started a vending machine company and Ruth helped. Every morning they stirred the thick sugar and water mixture that went into the then-high-tech coffee machines that dispensed cups and filled them with piping-hot coffee—and the sugar mixture if you pushed the right button. This company took off, and Adolph's Coffee Services is still run today by their son, Jay Teitelbaum.[6]

Ruth worked for Pollock Paper in their early computer division, and later for Chance Vought Aircraft, in their computing department. Chance Vought designed and built thousands of planes capable of launching off of and landing on Navy aircraft carriers during and after WWII. She worked until her second son, David, was born in 1954.

Thereafter, she raised her family and was deeply involved in managing the family finances. Her sons remember that she managed the family stock portfolio and did well. She read the *Wall Street Journal* every day and talked with Adolph every night about business and the New York Stock Exchange before they headed into dinner with the family.

Ruth's sons and daughters-in-law remember Ruth and Adolph as a close couple, a model for those who knew them, and were very close to the many friends that they made. They remember Ruth as generous and warm.

During the ENIAC patent trial in the 1970s, Ruth traveled to Arlington, Virginia, to give a deposition about her work on the ENIAC in the mid-1940s.

After Adolph retired and handed the reins to Jay, he and Ruth traveled around the country, visiting elder hostels. They liked to hike, see new places, and discover new things together.

Over the years, Ruth and Adolph kept in touch with Marlyn and Phil and got together when they came to Philadelphia to visit family. When Kay invited them to her reunion in 1985, they attended and it was a joyous event. It would also be the last time that the ENIAC Programmers would be together. Ruth died the following year, just before the fortieth anniversary of ENIAC.

When the ENIAC Programmers were inducted into the Hall of Fame by Women in Technology International in 1997, Adolph attended, wanting to celebrate Ruth's work and her legacy. He was the only man onstage.

Frances Bilas Spence (1922–2012)

Fran continued to work at Aberdeen after she was married, and it was easy to do so, as Homer Spence remained working with ENIAC for years to come. Fran became the mother of three sons, Joseph, Richard, and William, and enjoyed the roles of mother and wife. When Homer later took a job in Syosset, New York, the family moved there and set down new roots.

When Kay held her 1985 reunion, Fran and Homer were present and smiling in the photographs, but Fran did not attend later anniversaries and declined to participate in oral history projects. I understand that Homer was ill at or around the fiftieth anniversary of ENIAC in 1996, and this may have made participation difficult for Fran.[7] Fran passed away in 2012.

Marlyn Wescoff Meltzer (1922–2008)

After their marriage, Marlyn and Phil lived above Phil's dental office in Trenton, New Jersey, for the first decade of their marriage. Never one to shy away from a challenge, Marlyn became certified in taking dental X-rays and took the X-rays of patients for many years. She learned the other jobs, too, including mixing amalgams for fillings, serving as a hygienist, manning the phones, setting up appointments, and handling the billing and payments. Whatever the young dental practice needed, Marlyn learned it and did it.

In 1957, they bought a house in Yardley, Pennsylvania, with huge, bright windows, not far from the famous spot where General George Washington and his troops crossed the Delaware River during the Revolutionary War to win a key battle.

Marlyn and Philip raised two children, Joy and Hugh, and stayed in Yardley for the rest of their lives.

Over the years, Marlyn participated in many local organizations to help women train and find jobs and to raise money for families in need. She often served on their board of directors as treasurer and secretary, always willing to work hard for the mission of the nonprofit, but never seeking the roles of president or vice president because she did not like to give speeches.

Marlyn continued to critique her knitting skills but became

known at the local hospitals for the thousands of hats she knit for cancer patients, adults and children.

Marlyn and Phil traveled to many locations not then widely visited by Americans, including China, Hong Kong, Italy, Greece, and Thailand. By cruise ship, they went to the Soviet Union and St. Petersburg to see the great Hermitage, the world's second-largest art museum, first opened by Catherine the Great in 1764.

Israel was a frequent destination for them, and not solely for tourism. Phil and Marlyn adopted a kibbutz and visited it once a year to provide free dental services to its members. After Phil's retirement, they stayed for a month at a time, living on the kibbutz and working out of its small dental office.

In 1997, when Marlyn traveled to Santa Clara, California, for the WITI Hall of Fame induction, she was so pleased to meet thousands of young women working in technology and computing fields. "It was an eye-opener," she said, and "very exciting."[8]

Jean Jennings Bartik (1924–2011)

Like Betty, Jean joined John and Pres and worked for Eckert-Mauchly Computer Corporation. For two and a half years, she was a Programmer on BINAC and UNIVAC I. She then became a logic designer for UNIVAC I.

When Bill accepted a job in the Washington, DC, area, Jean found a position at Remington Rand's Washington, DC, offices, now merged with Eckert-Mauchly Computer Corporation. A few years later, they moved back to Philadelphia and Jean decided to stay home and raise the three children born there: Tim, Jane, and Mary. She loved being a mother.

Jean became active in the League of Women Voters and local

Children's Hospital, and returned to school at Penn, earning a master's degree in education in 1967. After her divorce, she returned to work, joining Isaac Auerbach of Auerbach Publications and eventually becoming editor of its *Auerbach Minicomputer Reports*, featuring the "minicomputers" of the day. Minicomputers were still large computers by today's standards, but "mini" in comparison to the computers that came before them, including ENIAC, and they put computer power within reach of a much wider set of smaller businesses and organizations.[9] (Most minicomputers cost between $10,000 and $25,000 in 1970, far cheaper than the mainframes of the day.)

Jean enjoyed working to help a new audience of computer users find the right computers for their needs. She stayed with Auerbach Publications for almost a decade and then joined high-tech companies, including Interdata, Honeywell, and Data Decisions.

Jean attended the WITI Hall of Fame. Soon afterward, a young cousin, one of the Jennings clan who attended the Northwest Missouri State University (NWMSU) nearby, told the university about Jean's accomplishments. Jon Rickman, Vice President of Information Systems at the school, contacted Jean and in July 2001, Jean's picture appeared on the front cover of *Northwest Alumni* magazine, with the title "Computer Pioneer, Jean Bartik '45, Opening the Door of Opportunity." A year later, Jean was honored as commencement speaker and proudly rode with the university's president in the commencement procession on April 27, 2002.

Jean thought carefully about her commencement address and titled it "10 Proverbs of Life." She gave her favorite pieces of advice to the young graduates, advice she learned from decades of work and experience, including:

#1 It pays to dream...

#3 When opportunity knocks, answer the door...

#5 New and exciting beats dull and boring...

She ended with a favorite phrase of hers: "You can get the girl out of Missouri but you can't get Missouri out of the girl," which she said with a huge belly laugh.[10]

NWMSU created a home for Jean's collections, files and memoirs, now part of the Jean Jennings Bartik Computing Museum. It houses Jean's treasured notes, diagrams, and photographs of ENIAC, BINAC, UNIVAC, minicomputers, and more.

Jon Rickman and Kim Todd of NWMSU helped Jean edit her manuscript, along with her son, Tim Bartik, and it is now published as the book *Pioneer Programmer: Jean Jennings Bartik and the Computer that Changed the World* (Truman State University Press). It is written in Jean's inimitable style, with her deep computer knowledge, candor, and strong sense of humor.

Frances Elizabeth "Betty" Snyder Holberton (1917–2001)

It is impossible to encapsulate Betty's full range of contributions to computing or programming in just a few paragraphs. She started in computing in the mid-1940s and continued with a career that ran uninterrupted into the early 1980s. During that time, she created cutting-edge tools for programming, including the first modern sort routine and a very early software application. She, Jean, and Grace Hopper joined John Mauchly and about two dozen early programming pioneers to sign the charter of the Association for Computing Machinery at Columbia University on September 15, 1947, and dedicate it to computing and programming education and research.

After her work for BRL at the Proving Ground, Betty joined Eckert-Mauchly Computer Corporation as one of the first ten employees. Her rank was considered "management," and her title was elevated to "engineer."

When Pres designed the magnetic tape and tape drives and needed an input/output system, he asked Betty to design and write one, and she did. It was very versatile and could store data and read it in forward or reverse order.

When John needed an instruction code for UNIVAC, he asked Betty to write it. The result was the C-10 instruction code, which shipped on every UNIVAC I. Betty's C-10 was so powerful, mnemonic, and easy to remember, said Mildred Koss, a UNIVAC Programmer for EMCC and later VP of information technology for Harvard University.

Betty then used the C-10 to create a sort routine and insisted it be shipped with every UNIVAC I. Her sort routine became the "killer app" of the 1950s.

In his book *The Art of Computer Programming*, Volume 3: *Sorting and Searching*, Professor Donald Knuth described Betty's groundbreaking sort routing as "a perfect overlap of reading, writing, and computing, using six input buffers." In her obituary in the *New York Times* in 2001, Knuth hailed Betty as "a real software pioneer."

Betty joined John Mauchly in traveling widely to help businesses think about how they could use computers. She talked with Prudential Insurance, overflowing with customer data including premiums and payments, and Nielsen, the national survey company, which could not process their surveys as quickly as they wanted. Both became early customers of UNIVAC.

In 1950, Betty married John Holberton, and they delayed their honeymoon for a year while Betty joined John and Pres in preparing the first UNIVAC I for shipment. A year later, the couple left for the United Kingdom by ship for their long-overdue honeymoon.

Betty played key roles in writing and testing new and powerful programming languages. She served as chief editor of the initial COBOL report (as editor of the language, she reviewed every command and specification of the new language before it was introduced) and wrote important test routines for Fortran, another powerful early programming language.

Most of the time, Betty was the only working mother she knew in Potomac, Maryland. When she arrived home after work, she started a second shift as the leader of her daughters' Girl Scouts troop and other activities. Betty and John's two daughters, Priscilla and Pamela, heard a lot of computer talk around the table, and Betty said that as young children, they ran around the yard yelling, "hardball, software, COBOL" as one of their playtime refrains.

In the 1960s, opposed to the Vietnam War, Betty and John both left military work. Betty joined the National Bureau of Standards, where she worked on many projects and served in many roles. When NATO needed an American expert to check computer installations in Italy, they sent Betty to figure out the problems. Later she was asked to serve on the US delegation to negotiate a European treaty with standards for magnetic tapes so that they could be passed back and forth across the Atlantic with ease. In these positions, Betty was one of very few women, which she found lonely. But it did not stop her from her doing work as the most senior computing member of both teams.

For her entire life, Betty cared deeply about computer users: "The user was the one who paid the bill," she said, "and so that was very important."[11]

Before Betty retired from the National Bureau of Standards, she received the Silver Medal for "exceptional service"—a rare honor. Secretary of Commerce Elliot Richardson awarded it to her, and as he handed her the medal, he leaned over and whispered in her ear, "I'm glad a woman made it." Her career spanned over forty years.

At eighty years old, and following a stroke, Betty traveled from Washington, DC, to Santa Clara, California, to join other members of the ENIAC 6 at the WITI induction ceremony for the Hall of Fame. The head of IBM's Mainframe Division, Linda Sanford, delivered a speech outlining the six women's work during World War II, calculating ballistics trajectories, and programming ENIAC. No one in the audience had previously known this story.

Betty gave a short acceptance speech. She made jokes about the Demonstration Day bug, calling it her first do-loop error, but then grew serious.

> We didn't have a thing like that [WITI] fifty years ago," she said, "and we could have sure used it. I think that all of you women will be successful if you can put people like that behind you. You just need it. And I am so proud of those people who recognize women.[12]

Her voice cracked and she began to cry. The applause spread through the convention room. Then the audience rose in a wave across the huge room. First the women in the front rose, and then the women behind them, all the way to the back, a massive wave. Everyone was on their feet, applauding and crying, giving Betty, Jean, Marlyn, and Adolph an unexpected standing ovation. The noise grew louder and louder until it filled the room.

Endnotes

Looking for Women Math Majors

1. Josephine Benson Manfredi, in interview with author and Amy Sohn, February 29, 2020.
2. "Chestnut Hill College Graduates Class of 107," *Philadelphia Inquirer*, June 3, 1942, 13.
3. "Chestnut Hill College Graduates Class of 107."
4. "Students Graduating into World at War," *Philadelphia Inquirer*, June 3, 1942, 20.
5. Josephine Betts Caldwell, "History of the Class of 1940," Record Book of the Class of 1940 University of Pennsylvania, 24, https://archives.upenn.edu/digitized-resources/docs-pubs/womens-yearbooks/yearbook-1940.
6. Kathleen "Kay" McNulty Mauchly Antonelli, interview by author and directed by David Roland, recorded in the home of Mrs. Antonelli, September 18, 1997, transcript, ENIAC Programmers Oral History Project, 4. (Henceforth cited as Antonelli, Oral History.)
7. "College Girls to Present Style Show," *Philadelphia Inquirer*, May 10, 1942, 68.
8. Antonelli, Oral History, 5.
9. See, e.g., *Philadelphia Inquirer*, June 28, 1942, 53.
10. "The Kathleen McNulty Mauchly Antonelli Story," March 26, 2004, https://sites.google.com/a/opgate.com/eniac/Home/kay-mcnulty-mauchly-antonelli. (Henceforth cited as Antonelli, "The KMMA Story.")
11. Kimberly Amadeo, "Unemployment Rate by Year Since 1929 Compared to Inflation and GDP," The Balance, November 10, 2021, https://www.thebalance.com/unemployment-rate-by-year-3305506.
12. For example, "Women & World War II," Metropolitan State University of Denver, https://temp.msudenver.edu/camphale/thewomensarmycorps/womenwwii/.

13. For example, "'Rosie the Riveter' Song Lyrics," http://jackiewhiting.net/us/rosielyrics.html.
14. Alfred Palmer, photographer, *The More Women at Work, the Sooner We Win!*, 1943, World War II Posters. Prints and Photographs Division, LC-USZC4-5600, https://memory.loc.gov/ammem/awhhtml/awpnp6/d13.html.
15. Norman Rockwell Museum; "Rosie The Riveter—1943," https://www.nrm.org/rosie-the-riveter/.
16. Ruth Milkman, "Redefining 'Women's Work': The Sexual Division of Labor in the Auto Industry during World War II," in "Women and Work," *Feminist Studies* 8, no. 2 (Summer 1982): 336–72.
17. Evelyn Steele, Wartime Opportunities for Women (New York, 1943), 99–100, cited in Jennifer S. Light, "When Computers Were Women," *Technology and Culture* 40, no. 3 (July 1999): 457.
18. "Specialized War Jobs Seek Girl Math Majors," *Brooklyn Daily Eagle*, January 28, 1943, 4.
19. "'Haven't Felt Pinch of War,' Women Told," *Philadelphia Inquirer*, February 11, 1943, 20.
20. Antonelli, Oral History, 5.
21. Antonelli, "The KMMA Story."
22. Antonelli, "The KMMA Story."
23. Antonelli, "The KMMA Story."
24. Antonelli, "The KMMA Story."
25. Antonelli, "The KMMA Story."
26. "We spoke only Gaellic in our house in Ireland and the United States," Kay wrote in her autobiographical piece. Antonelli, "The KMMA Story."
27. Antonelli, "The KMMA Story."
28. Antonelli, "The KMMA Story."
29. Antonelli, Oral History, 2.
30. Antonelli, Oral History, 2.
31. Antonelli, Oral History, 2.
32. Antonelli, "The KMMA Story."
33. Antonelli, "The KMMA Story."
34. Antonelli, "The KMMA Story."
35. Antonelli, "The KMMA Story."
36. Antonelli, "The KMMA Story."
37. Antonelli, Oral History, 4.
38. W. Barkley Fritz, "The Women of ENIAC," *IEEE Annals of the History of Computing* 18, no. 3 (1996): 23.
39. Engineering and Technology Wiki, "Frances Spence," https://ethw.org/Frances_Spence.

40. Antonelli, in interview with author, April 18, 2000.
41. Engineering and Technology Wiki, "Frances Spence."
42. Manfredi, interview, February 29, 2020.
43. Antonelli, "The KMMA Story."
44. "Changes in Women's Occupations 1940–1950," *Women's Bureau Bulletin 253*, United States Department of Labor (1954): 3.
45. Antonelli, "The KMMA Story."
46. Antonelli, "The KMMA Story."
47. Herbert Ershkowitz, "World War II," Encyclopedia of Greater Philadelphia, 2011, https://philadelphiaencyclopedia.org/archive/world-war-ii/.
48. Ershkowitz, "World War II."
49. Manfredi, interview, February 29, 2020. Manfredi described her friendship with Kathleen McNulty and her own activities in college and during WWII. She shared information about the Cinderella Group of approximately fifty young women who hosted dances for soldiers during WWII that ended promptly at midnight.
50. See, e.g., "Edward R. Murrow, A Hero of Broadcast Journalism," https://www.dmagazine.com/publications/d-magazine/1990/august/edward-r-murrow-a-hero-of-broadcast-journalism/.
51. Richard Holmes, "Maria Mitchell at 200: A Pioneering Astronomer Who Fought for Women in Science," *Nature*, June 18, 2018, https://www.nature.com/articles/d41586-018-05458-6.
52. Antonelli, Oral History, 5.
53. Professional and scientific jobs were "based on the established principles of a profession or science" and required professional, scientific, or technical training equivalent to recognized-standing college or university training. Rachel Fesler Nyswander and Janet M. Hooks, *Employment of Women in the Federal Government, 1923 to 1939, Bulletin of the Women's Bureau No. 182*, US Government Printing Office, 1941, 12.
54. W. Barkley Fritz, "ENIAC—A Problem Solver," *IEEE Annals of the History of Computing* 16, no. 1 (1994): 28.
55. Frances Elizabeth "Betty" Snyder Holberton, interview by author and directed by David Roland, recorded in the library of the Shady Grove Center, Rockville, MD, September 23–24, 1997, transcript, ENIAC Programmers Oral History Project, 12. (Henceforth cited as Holberton, Oral History.)

We Were Strangers There

1. Antonelli, Oral History, 9.
2. J. N. Shurkin, *Engines of the Mind: The Evolution of the Computer from Mainframes*

to Microprocessors (New York: W. W. Norton & Company, 1996):118–19. (Henceforth cited as Shurkin, *Engines of the Mind.*)

3. Peter Eckstein, "J. Presper Eckert," *IEEE Annals of the History of Computing* 18, no. 1 (1996): 36.
4. Harold Pender to Dr. Musser, June 8, 1942, University Relations Information Files, UPF 5I, Box 108, File "Ballistics Calculations Courses," University Archives and Records Center, University of Pennsylvania.
5. Antonelli, Oral History, 5.
6. Antonelli, Oral History, 5.
7. Antonelli, Oral History, 5.
8. Antonelli, Oral History, 5.
9. Fritz, "The Women of ENIAC," 13–28.
10. Thomas J. Bergin, ed., *50 Years of Army Computing, From ENIAC to MSRC: A Record of a Symposium and Celebration, November 13 and 14, 1996* (sponsored by the Army Research Laboratory and U.S. Army Ordnance Center & School, September 2000), 40.
11. Fritz, "The Women of ENIAC," 15.
12. "It's quite complex. It depends on the barometric pressure, the humidity, the curvature of the earth, all kinds of factors. It's a very complicated equation that has to be solved for each data point, for each range that you want to compute the aiming of the gun at," Dr. Paul Ceruzzi, Historian, Smithsonian Institution. Kathy Kleiman, *The Computers: The Remarkable Story of the ENIAC Programmers*, produced by Kathy Kleiman, Jon Palfreman, and Kate McMahon, video documentary, Women Make Movies distributor, 2014.
13. Antonelli, Oral History, 6.
14. Discussion of Shirley Blumberg Melvin about problems with the Monroe and Marchant desktop calculators that she worked with. LeAnn Erickson, director, *Top Secret Rosies*, 2011, https://www.amazon.com/Top-Secret-Rosies-Female-Computers/dp/B00443FMKC.
15. Antonelli, Oral History, 9.

Nestled in a Corner of the Base

1. Aberdeen Proving Ground, "History," https://home.army.mil/apg/index.php/about/history.
2. Aberdeen Proving Ground, "History."
3. See, e.g., Shelford Bidwell, ed., *Brassey's Artillery of the World: Guns, Howitzers, Mortars, Guided Weapons, Rockets and Ancillary Equipment in Service with the Regular and Reserve Forces of All Nations* (London: Brassey's Publishers Ltd., 1977), 29.
4. Communications methods of World War I included field telephones, radios,

messenger dogs, and even carrier pigeons. In one famous example, "Cher Ami, a carrier pigeon of the US Army Signal Corps flew wounded to carry the message of an isolated battalion in need of help. The lives of almost two hundred soldiers were saved." Smithsonian National Museum of American History, "Cher Ami," https://www.si.edu/object/cher-ami%3Anmah_425415.

5. Saunders Mac Lane, *Oswald Veblen, 1880–1960*, National Academy of Sciences, http://www.nasonline.org/publications/biographical-memoirs/memoir-pdfs/veblen-oswald.pdf.
6. The Range Firing Section, under Major Oswald Veblen, was one of the original nine divisions of Aberdeen Providing Ground. Henry Reid, "Ballisticians in War and Peace, Volume I, 1914–1956." U.S. Army Research Laboratory, Aberdeen Proving Ground, MD, https://apps.dtic.mil/sti/pdfs/ADA300523.pdf.
7. Reid, "Ballisticians in War and Peace, Volume I," 3.
8. Reid, "Ballisticians in War and Peace, Volume I," 4–5.
9. Herman H. Goldstine, *The Computer from Pascal to Von Neumann* (Princeton, NJ: Princeton University Press, 1993), 132.
10. Reid, "Ballisticians in War and Peace, Volume I," 8.
11. For example, 155 mm M59 Long Tom, WeaponSystems.net, https://weaponsystems.net/system/920-155mm+M59+Long+Tom.
12. Reid, "Ballisticians in War and Peace, Volume I," 9.
13. BRL's Scientific Advisory Committee, 1940, First Meeting, https://ftp.arl.army.mil/~mike/comphist/40sac/index.html.
14. Reid, "Ballisticians in War and Peace, Volume I," 29.
15. Goldstine, *The Computer from Pascal to Von Neumann*, 132.

Give Other People as Much Credit as You Give Yourself

1. Holberton, Oral History, 30.
2. See, e.g., "In the Military during World War II, National Women's History Museum," https://www.womenshistory.org/resources/general/military.
3. Holberton, Oral History, 30.
4. Holberton, Oral History, 43.
5. Holberton, Oral History, 46.
6. Caldwell, "History of the Class of 1940," 21–23, https://archives.upenn.edu/digitized-resources/docs-pubs/womens-yearbooks/yearbook-1940.
7. Holberton, Oral History, 45.
8. Farm Journal Magazine, https://www.agweb.com/farm-journal-magazine.
9. Holberton, Oral History, 46.
10. Fritz, "The Women of ENIAC," 17.
11. Holberton, Oral History, 31.

12. David Alan Grier, *When Computers Were Human* (Princeton, NJ: Princeton University Press, 2005), 260.

We Found Things in a Not Very Good State

1. Goldstine, *The Computer from Pascal to Von Neumann*, 133.
2. Seabright McCabe, "Adele Goldstine, The Woman Who Wrote the Book," *SWE Magazine*, spring 2019, https://alltogether.swe.org/2019/05/adele-goldstine-the-woman-who-wrote-the-book/.
3. Goldstine, *The Computer from Pascal to Von Neumann*, 134.
4. Jean Jennings Bartik, interview by author and directed by David Roland, recorded in the home of Ms. Bartik, September 17, 1997, transcript, ENIAC Programmers Oral History Project, 10. (Henceforth cited as Bartik, Oral History.)
5. Citing an entry in Adele Goldstine's notebook, circa 1962. Bergin, *50 Years of Army Computing*, 28.

Adding Machines and Radar

1. Marlyn Wescoff Meltzer, interview by author and directed by David Roland, recorded in the home of Mrs. Meltzer, September 16, 1997, transcript, ENIAC Programmers Oral History Project, 6. (Henceforth cited as Meltzer, Oral History.)
2. Meltzer, Oral History, 7.
3. Meltzer, Oral History, 5, and discussion with author. Also, "Houses with Family Members Born in Italy, West Philadelphia 1940," West Philadelphia Collaborative History, https://collaborativehistory.gse.upenn.edu/media/houses-family-members-born-italy-west-philadelphia-1940.
4. Joy Meltzer (Meltzer's daughter), in interview with author, April 29, 2021.
5. Meltzer, Oral History, 1.
6. Meltzer, in interview with Thomas Petzinger Jr., 1996.
7. Meltzer, Oral History, 3.
8. Meltzer, Oral History, 9.
9. Meltzer, Oral History, 6.
10. Antonelli, Oral History, 12.
11. Holberton, Oral History, 9, and LOOSE LIPS MIGHT SINK SHIPS poster (e.g., https://www.nh.gov/nhsl/ww2/loose.html).
12. Meltzer, in interview with Thomas Petzinger Jr., 1996.
13. Meltzer, Oral History, 7–8.
14. "Anecdotes from Some of the Pioneers, from Joseph Chapline," VIP Club, established in 1980: Information Technology Pioneers: Retirees and Former

Employees of Unisys, Lockheed Martin, and Their Predecessor Companies, Chapter 34, Blue Bell, http://vipclubmn.org/BlueBell.html. (Henceforth cited as "Anecdotes from some of the Pioneers.")

15. Antonelli, "The KMMA Story."
16. John Costello, "As the Twig Is Bent: The Early Life of Mauchly," *IEEE Annals of the History of Computing* 18, no. 1 (1996): 50.
17. Antonelli, "The KMMA Story."
18. Antonelli, "The KMMA Story."
19. Antonelli, "The KMMA Story."
20. Paul David, director, *Mauchly: The Computer and the Skateboard*, Blastoff Media, 2011.

3436 Walnut Street

1. Meltzer's handwritten notes from Adele Goldstine's class written on the back of an AAUW flier (likely extra note paper available in the classroom), which announces Adele's visit to AAUW headquarters in New York City to recruit college women. Copy given to author by Mrs. Meltzer.
2. Meltzer, Oral History, 14, 23.
3. Meltzer, in interview with Thomas Petzinger Jr., 1996.
4. Meltzer, Oral History, 11.
5. Meltzer, Oral History, 11.
6. Meltzer, Oral History, 9.
7. Priscilla Holberton, in interview with author, 2003.
8. Meltzer, Oral History, 13.
9. Meltzer, Oral History, 13.
10. Meltzer, Oral History, 15.
11. Erickson, *Top Secret Rosies*.
12. Grace Yeager Potts Vaughan obituary, https://www.legacy.com/obituaries/timesunion/obituary.aspx?n=grace-yeager-potts-vaughan&pid=425473.
13. Yvonne Latty, "Alyce McLaine Hall, Talented in Math," *Philadelphia Daily News*, November 20, 2003, 61.
14. Yvonne Latty, "Alyce McLaine Hall, Talented in Math."
15. Seabright McCabe, "Finding Alyce Hall," *SWE Magazine*, Winter 2014, 28.
16. Meltzer, Oral History, 11.
17. Scott McCartney, *ENIAC: The Triumphs and Tragedies of the World's First Computer* (New York: Walker and Company, 1999), 54.
18. John Mauchly, in interview with Nancy Stern, Oral History, May 6, 1977.
19. Grier, *When Computers Were Human*, 260, and American Association of

University Women flyer announcing Adele Goldstine's upcoming trip and college visits to New York City. Shared by Meltzer with author.

20. "Army Needs Math Majors," *Brooklyn Daily Eagle*, June 14, 1943, 4.
21. "Army Needs Math Majors."
22. Grier, *When Computers Were Human*, 260.
23. Ruth Teitelbaum, in interview with family, 1986.
24. Teitelbaum, in interview with family, 1986.
25. Teitelbaum, in interview with family, 1986.
26. Teitelbaum, in interview with family, 1986.
27. Teitelbaum, in interview with family, 1986.
28. Teitelbaum, in interview with family, 1986.
29. Teitelbaum, in interview with family, 1986.
30. Typed sheets of these lyrics shown in Erickson, *Top Secret Rosies*.
31. Erickson, *Top Secret Rosies*.

The Monster in the Basement

1. Antonelli, Oral History, 7.
2. Antonelli, Oral History, 7.
3. Shurkin, *Engines of the Mind*, 97.
4. Shurkin, *Engines of the Mind*, 101.
5. Fritz, "The Women of ENIAC," 16.
6. Antonelli, Oral History, 7.
7. Professor Joseph Weizenbaum, MIT, in interview with author, 1986–1987.
8. Antonelli, Oral History, 8.
9. Antonelli, Oral History, 8.
10. Antonelli, Oral History, 8.
11. "Anecdotes from Some of the Pioneers."
12. Kathleen McNulty Mauchly Antonelli, "Luncheon Speech, Reminiscences," introduction speech at the 40th anniversary of ENIAC, Transcript by author, October 1986.
13. Fritz, "The Women of ENIAC," 14.
14. Antonelli, Oral History, 9.
15. David, *Mauchly: The Computer and the Skateboard*, Blastoff Media, 2001.

The Lost Memo

1. Shurkin, *Engines of the Mind*, 134.
2. Goldstine shared his reason for optimism: If the Army Ordnance Department was capable of giving General Motors $1 million during wartime to build a prototype tank and then discard the tank if it did not suit their needs, "why

not spend equal or similar amounts of money on trying out an electronic computer." Shurkin, *Engines of the Mind*, 137.

3. John Mauchly, in interview with Nancy Stern, May 6, 1977.
4. John Mauchly, in interview with Nancy Stern, May 6, 1977.
5. Shurkin, *Engines of the Mind*, 134.
6. Dorothy Shisler, Ursinus College 1941, *U.S., School Yearbooks, 1900–1999* [database online]. Provo, UT, USA: Ancestry.com Operations, 2011.
7. "Anecdotes from some of the Pioneers," and John Mauchly, in interview with Nancy Stern, May 6, 1977.
8. Fritz, "ENIAC—A Problem Solver," 25–45.
9. Kathleen R. Mauchly, "John Mauchly's Early Years," *Annals of the History of Computing* 6, no. 2 (April 1984): 137.
10. Shurkin, *Engines of the Mind*, 137.
11. Shurkin, *Engines of the Mind*, 110.
12. Mauchly, "John Mauchly's Early Years," 137.
13. McCartney, *ENIAC*, 71.
14. John Ambrose Fleming invented diodes in 1904. ENIAC would use a more advanced version, a triode, invented by Lee De Forest in 1906.
15. Mauchly, "John Mauchly's Early Years," 137.
16. Shurkin, *Engines of the Mind*, 150–152.
17. Eckstein, "J. Presper Eckert," 25–44.
18. Eckstein, "J. Presper Eckert," 29.
19. Shurkin, *Engines of the Mind*, 122.
20. Eckstein, "J. Presper Eckert," 37–38, citing J. Presper Eckert Jr., "Development of the ENIAC: Session One," in interview with David Allison, Smithsonian Video History Program, February 2, 1988.
21. Mauchly, "John Mauchly's Early Years," 131.
22. Shurkin, *Engines of the Mind*, 123.
23. Shurkin, *Engines of the Mind*, 166–67.
24. Shurkin, *Engines of the Mind*, 166–67.
25. Shurkin, *Engines of the Mind*, 133.
26. Eckstein, "J. Presper Eckert," 39.
27. Eckstein, "J. Presper Eckert," 19.

"Give Goldstine the Money"

1. John Mauchly wrote in his diary that Brainerd "was pooh-poohing the idea that anyone could ever seriously consider such a thing." McCartney, *ENIAC*, 57. J. Presper Eckert, in interview with Nancy Stern, October 28, 1977.
2. Bergin, *50 Years of Army Computing*, 30.

3. Mauchly, "John Mauchly's Early Years," 137.
4. Goldstine, *The Computer from Pascal to Von Neumann*, 149.
5. Goldstine, *The Computer from Pascal to Von Neumann*, 150.
6. McCartney, *ENIAC*, 57.
7. Shurkin, *Engines of the Mind*, 137.
8. Shurkin, *Engines of the Mind*, 137.
9. John Mauchly, in interview with Esther Carr, 1977.
10. John Mauchly, in interview with Esther Carr, 1977.

Dark Days of the War

1. Antonelli, "The KMMA Story."
2. Antonelli, "The KMMA Story."
3. Antonelli, "The KMMA Story."
4. Shurkin, *Engines of the Mind*, 148.
5. Shurkin, *Engines of the Mind*, 148.
6. Shurkin, *Engines of the Mind*, 149.
7. Shurkin, *Engines of the Mind*, 148.
8. McCartney, *ENIAC*, 65–66.
9. Meltzer, Oral History, 9.
10. Marlyn Wescoff Meltzer, in interview with author, February 6, 1996.
11. Meltzer, in interview with author, February 6, 1996.

"All That Machinery Just to Do One Little Thing Like That"

1. McCartney, *ENIAC*, 87.
2. McCartney, *ENIAC*, 69.
3. Arthur and Alice Burks, in interview with Nancy Stern, 1980.
4. Antonelli, "The KMMA Story."
5. McCartney, *ENIAC*, 77. We note that in chapter 3, note 15, of *ENIAC in Action*, Thomas Haigh, Mark Priestley, and Crispin Rope list the names of about three dozen women who they found in accounting statements of this period with a note lamenting these women being "literal footnotes" to the ENIAC's history. We hope they and other computer historians will find the families of these women and learn more about their WWII stories. We note that the history of women in computing is best told by both women and men.
6. Pine Camp, now Fort Drum, in the 1930s and '40s, https://www.northcountryatwork.org/collections/pine-camp-now-fort-drum-in-the-1930s-and-40s/.
7. Jean J. Bartik and Frances E. (Betty) Snyder Holberton, interview by Henry S. Tropp, transcript, Computer Oral History Collection, 1969–1973, 1977. Archives

Center, Smithsonian National Museum of American History. (Henceforth cited as Smithsonian Oral History of Bartik and Holberton.)

8. Mike Strasser, Fort Drum exhibit to highlight sonic deception training at Pine Camp during WWII, https://www.army.mil/article/240486/fort_drum_exhibit_to_highlight_sonic_deception_training_at_pine_camp_during_wwii. (Henceforth cited as Army Pine Camp website).
9. Army Pine Camp website.
10. Smithsonian Oral History of Bartik and Holberton, 1973.
11. Antonelli, "Luncheon Speech, Reminiscences," 1986.
12. Antonelli, "Luncheon Speech, Reminiscences," 1986.
13. Antonelli, Oral History, 10
14. Antonelli, in interview with author, July 20, 2003.
15. Antonelli, Oral History, 10.
16. Antonelli, Oral History, 11.
17. Antonelli, Oral History, 11.
18. Antonelli, Oral History, 11.
19. Antonelli, Oral History, 11.

The Kissing Bridge

1. Bartik, Oral History, 5.
2. Bartik, Oral History, 5.
3. Bartik, Oral History, 6.
4. Bartik, Oral History, 4.
5. Bartik, Oral History, 1.
6. Bartik, Oral History, 2.
7. Bartik, Oral History, 1.
8. Jean Jennings Bartik, *Pioneer Programmer: Jean Jennings Bartik and the Computer that Changed the World* (Kirksville, MO: Truman State University Press, 2013), 33.
9. Bartik, *Pioneer Programmer*, 30.
10. Bartik, Oral History, 1.
11. Bartik, Oral History, 3.
12. Bartik, *Pioneer Programmer*, 35.
13. Bartik, Oral History, 2–3.
14. Bartik, Oral History, 4.
15. Bartik, *Pioneer Programmer*, 44.
16. Bartik, *Pioneer Programmer*, 44.
17. Bartik, *Pioneer Programmer*, 46.
18. Bartik, *Pioneer Programmer*, 48.
19. Bartik, *Pioneer Programmer*, 51.

20. Bartik, *Pioneer Programmer*, 50.
21. Bartik, *Pioneer Programmer*, 51.
22. Bartik, *Pioneer Programmer*, 51–52.
23. Bartik, *Pioneer Programmer*, 53.
24. Bartik, *Pioneer Programmer*, 55.
25. Bartik, Oral History, 6.
26. Bartik, *Pioneer Programmer*, 58.
27. Bartik, Oral History, 6.
28. Bartik, *Pioneer Programmer*, 58.
29. Bartik, *Pioneer Programmer*, 58.
30. Bartik, Oral History, 6.

Are You Scared of Electricity?

1. Bartik, *Pioneer Programmer*, 58.
2. Bartik, *Pioneer Programmer*, 58.
3. Bartik, *Pioneer Programmer*, 58–59.
4. Bartik, *Pioneer Programmer*, 59.
5. "War Department, Notification of Personnel Action (field), TO: Miss Betty J. Jennings, THROUGH: Lt. Landry, Ballistics Research Laboratory Division." March 30, 1945. Copy shared with author.
6. Bartik, *Pioneer Programmer*, 60.
7. Bartik, *Pioneer Programmer*, 59.
8. Bartik, *Pioneer Programmer*, 59.
9. Bartik, *Pioneer Programmer*, 60.
10. Bartik, Oral History, 11.
11. Bartik, *Pioneer Programmer*, 60.
12. Bartik, *Pioneer Programmer*, 61.
13. "Roosevelt Dead," *Philadelphia Inquirer*, April 13, 1945, 1.
14. Bartik, *Pioneer Programmer*, 61.
15. Sarah Pruitt, "How FDR's 'Fireside Chats' Helped Calm a Nation in Crisis," History.com, April, 7, 2020, https://www.history.com/news/fdr-fireside-chats-great-depression-world-war-ii.
16. Pruitt, "How FDR's 'Fireside Chats' Helped Calm a Nation in Crisis."
17. Bartik, *Pioneer Programmer*, 65.
18. Bartik, Oral History, 7.
19. Bartik, Oral History, 7.
20. Bartik, Oral History, 7.
21. Bartik, *Pioneer Programmer*, 65.

Learning It Her Way

1. Antonelli, Oral History, 10.
2. Antonelli, Oral History, 10.
3. Antonelli, "The KMMA Story."
4. Fritz, "The Women of ENIAC," 15.
5. Goldstine, *The Computer from Pascal to Von Neumann*, 202.
6. Meltzer, Oral History, 14.
7. Meltzer, Oral History, 14.
8. Holberton, Oral History, 1.
9. Antonelli, Oral History, 11.
10. Teitelbaum, in interview with family, 1986.
11. See, e.g., Aberdeen Proving Ground, "History," https://home.army.mil/apg/index.php/about/history.
12. Antonelli, Oral History, 24.
13. Antonelli, Oral History, 12, and Holberton, Oral History, 2.
14. See generally, IBM Control Boards (formerly called *plugboards*), http://www.columbia.edu/cu/computinghistory/plugboard.html.
15. Antonelli, Oral History, 12.
16. Holberton, Oral History, 2.
17. Holberton, Oral History, 2.
18. Holberton, Oral History, 2.
19. Teitelbaum, in interview with family, 1986.
20. Bartik, *Pioneer Programmer*, 68.
21. Holberton, Oral History, 2.

Surrounded by Vultures

1. Bartik, *Pioneer Programmer*, 69.
2. Bartik, *Pioneer Programmer*, 70.
3. Bartik in interview with author.
4. Bartik, *Pioneer Programmer*, 66.
5. Bartik, Oral History, 8.
6. Holberton, Oral History, 12.
7. Antonelli, Oral History, 24.
8. Bartik, *Pioneer Programmer*, 72.
9. Bartik, *Pioneer Programmer*, 72.
10. Bartik, *Pioneer Programmer*, 73.
11. Bartik, *Pioneer Programmer*, 74.

The Dean's Antechamber

1. Bartik, Oral History, 42.
2. Antonelli, Oral History, 13.
3. "More Jap Cities Put on 'Death List,'" *Philadelphia Inquirer*, August 1, 1945, 1. The August 1st paper stated that Major General Curtis LeMay gave twelve Japanese cities advanced notice and ordered them to evacuate their citizens before US bombers arrived.
4. By the end of July, US forces were shelling the main island of Honshu. A July 31, 1945, *Philadelphia Inquirer* front-page headline announced, "Fleet Shells Town 80 miles from Tokio (sic); Long Stretch of Coast Set Afire by Flames."
5. "12 More Jap Cities Placed on 'Surrender or Die' List," *Philadelphia Inquirer*, August 5, 1945, 1.
6. Diagram in newspaper showed the positions of US Army, Navy, and Air Force fleets and forces. "Where MacArthur Is Forging Mighty Invasion Forces," *Philadelphia Inquirer*, August 5, 1945, 2.
7. "12 More Jap Cities Placed on 'Surrender or Die' List," 1.
8. See, e.g., "'Hell to Pay' Sheds New Light on A-Bomb Decision," NPR, https://www.npr.org/templates/story/story.php?storyId=122591119.
9. "Atomic Bomb, World's Most Deadly, Blasts Japan; New Era in Warfare Is Opened by U.S. Secret Weapon," *Philadelphia Inquirer*, August 7, 1945, 1.
10. "Atom Bomb Hits Nagasaki," *Philadelphia Inquirer*, August 9, 1945, 1.
11. Antonelli, Oral History, 13.
12. Bartik, *Pioneer Programmer*, 83.
13. Emperor Hirohito, Accepting the Potsdam Declaration, Radio Broadcast, https://www.mtholyoke.edu/acad/intrel/hirohito.htm.
14. "PEACE," *Philadelphia Inquirer*, August 15, 1945, 1.
15. Antonelli, "The KMMA Story."
16. Fritz, "The Women of ENIAC," 14, 26.
17. Teitelbaum, in interview with family, 1986.

A New Project

1. Bartik, Oral History, 11. "And of course the prime problem was the trajectory. And the reason it was the prime problem was because that was what the machine [ENIAC] was financed for Aberdeen Proving Ground. So the acceptance test was going to be calculating a trajectory... [and] one of our missions was to program a trajectory for the acceptance test."
2. Antonelli, Oral History, 23.
3. Citing Ernie Pyle, *Right of the Line: A History of the American Field Artillery—US Army Field Artillery School, Fort Sill, Oklahoma, "Quotations," April 1984.*

4. Adele Goldstine, "Report on THE ENIAC (Electronic Numerical Integrator and Computer)," *Technical Report I, Volume I* (June 1, 1946), developed under the supervision of the Ordnance Department, United States Army, https://ftp.arl.army.mil/~mike/comphist/46eniac-report/index.html. (Henceforth cited as A. Goldstine, ENIAC *Technical Report I, Volume I.*)
5. Bartik, Oral History, 9.
6. Antonelli, Oral History, 13.
7. Holberton, Oral History, 2.
8. Holberton, Oral History, 3.
9. Antonelli, Oral History, 14.
10. Antonelli, Oral History, 20.
11. There are some discussions in other ENIAC sources that Arthur Burks worked on a trajectory program before ENIAC was built. If so, it may have helped to determine the size and scope of ENIAC, which expanded in a second agreement between the Moore School and the Army. But there is no account that Arthur shared any of his trajectory work with the ENIAC Programmers and no materials to show the women knew anything about it. What we do know is that the ENIAC Programmers were told it was their job to prepare the "acceptance program" for BRL, and they started their work with wiring, block and logical diagrams that Arthur provided them.
12. Antonelli, Oral History, 16. Betty echoed the thought: "We learned the equipment by reading the block diagrams . . . We never had any manual at all to read." Holberton, Oral History, 5.
13. Bartik, Oral History, 9.
14. Meltzer, Oral History, 15.

Divide and Conquer

1. Holberton, Oral History, 2.
2. Bartik, Oral History, 9.
3. Antonelli, "The KMMA Story."
4. ENIAC Patent 3,120,606, Sheet 37 of 91, high speed multiplier, https://patents.google.com/patent/US3120606A/en.
5. Meltzer, Oral History, 21.
6. Holberton, Oral History, 9.
7. ENIAC Patent 3,120,606, Sheet 25 of 91, accumulator, https://patents.google.com/patent/US3120606A/en.
8. Bartik, Oral History, 75.
9. Bartik, Oral History, 75.
10. Bartik, *Pioneer Programmer*, 75.

11. Bartik, Oral History, 9.
12. Bartik, *Pioneer Programmer*, 75.
13. Bartik, *Pioneer Programmer*, 88–89.
14. Meltzer, Oral History, 23.
15. Holberton, Oral History, 2.
16. Adele Goldstine's ENIAC *Technical Report I, Volume I*, would not be available to the ENIAC Programmers until 1947, two years after their ENIAC trajectory work and thus could not help them to learn the units of ENIAC. However, seventy-five years later, the *Technical Report* is a key resource for us, including for Adele's clear explanation of the four types of units of ENIAC. A. Goldstine, ENIAC *Technical Report I, Volume I*, p. I-2.
17. Adele Goldstine's ENIAC *Technical Report* also describes the cycling unit and its "fundamental train of signals repeated every addition time..." A. Goldstine, ENIAC *Technical Report I, Volume I*, p. I-3.

A Sequencing of the Problem

1. Adele Goldstine's ENIAC *Technical Report I, Volume I*, is dated the following year, June 1, 1946. A. Goldstine, *Technical Report I, Volume I*, title page.
2. Holberton, Oral History, 3.
3. Antonelli, Oral History, 14.
4. Holberton, Oral History, 6.
5. Thomas Petzinger, Jr., "History of Software Begins with the Work of Some Brainy Women," *Wall Street Journal*, November 15, 1996. This article was followed a week later by a second part written by the same author. Petzinger, "Female Pioneers Fostered Practicality in the Computer Industry," *Wall Street Journal*, November 22, 1996.
6. R. F. Clippinger, "A Logical Coding System Applied to the ENIAC," Report No. 673, Ballistic Research Laboratories, Aberdeen, MD, September 29, 1948, 3.
7. Fritz defined "direct programming" as "programming by directly connecting individual units with cables and setting switches to control the sequence of its [ENIAC's] operations needed to solve the specific problem at hand." Fritz, "ENIAC—A Problem Solver," 25.
8. Professor Brian Stuart, of the Drexel University Computer Science Department, has written extensively about ENIAC programming and debugging. His articles include: "Programming the ENIAC [Scanning Our Past]," *Proceedings of the IEEE* 106, no. 9 (2018): 1760–70; "Debugging the ENIAC [Scanning Our Past]," *Proceedings of the IEEE* 106, no. 12 (2018): 2331–45; and "Simulating the ENIAC [Scanning Our Past]," *Proceedings of the IEEE* 106, no. 4 (2018): 761–72.
9. Bartik, Oral History, 11, 19. Years later, Jean often would proclaim that "ENIAC was a son of a bitch to program!"

10. Bartik, Oral History, 20.
11. "Kay Mauchly on Finding Out about ENIAC, Programming It, and Marrying John Mauchly," Open Transcripts, January 1, 1977, http://opentranscripts.org/transcript/kay-mauchly-eniac-programming/.

A Tremendously Big Thing

1. Meltzer, Oral History, 16.
2. Bartik, *Pioneer Programmer*, 84.
3. Holberton, Oral History, 4.
4. Teitelbaum, in interview with family, 1986.
5. Antonelli, Oral History, 17.
6. "Deposition of Ruth Teitelbaum," ENIAC Patent Trial Collection, Box 16, University of Pennsylvania Archives, Deposition, 30. (Henceforth cited as Teitelbaum, Deposition.)
7. Teitelbaum, Deposition, 30.
8. For more detailed information on this setup, Jean explained: "Thus Accumulator 1 would receive the contents of Accumulator 2 from the Alpha memory bus and add it to its own contents. Pulse A-1 would start the next operation." Bartik, *Pioneer Programmer*, 84.
9. Bartik, Oral History, 13.
10. Bartik, Oral History, 12.
11. Bartik, *Pioneer Programmer*, 84.
12. Holberton, Oral History, 4.
13. "One million IBM cards were used." Shurkin, *Engines of the Mind*, 189.
14. Shurkin, *Engines of the Mind*, 189.
15. Fritz, "The Women of ENIAC," 29.
16. Teitelbaum, Deposition, 31.
17. McCartney, *ENIAC*, 104.

Programs and Pedaling Sheets

1. Holberton, Oral History, 3.
2. Because the BRL's trajectory program was classified, the ENIAC 6 could not keep a copy of their pedaling sheets. However, Kay kept a copy of the pedaling sheets of her next program, the Hartree problem described later in "Hundred-Year Problems & Programmers Needed," and shared them with the author. They are large white sheets titled "Compressible Boundary Layer, Zero-Order Functions" and use the pedaling sheet structure created by Betty and Jean.
3. Smithsonian Oral History of Bartik and Holberton, 1973.

4. "In modern terms, it [ENIAC] is a dataflow machine. The completion of an operation generates a control signal that is used to initiate the next operation in the sequence." Brian Stuart, "Debugging the ENIAC [Scanning Our Past]," *Proceedings of the IEEE* 106, no. 12, (2018): 2332.
5. Holberton, Oral History, 3.
6. Holberton, Oral History, 3–4.
7. Bartik, Oral History, 10.
8. Bartik, Oral History, 19, and Bartik, *Pioneer Programmer*, 84.
9. Holberton, Oral History, 2.
10. Holberton, Oral History, 2.
11. Holberton, Oral History, 2.
12. For example, a K-12 programming curriculum today teaches "16.2 If-Then Statements," cK-12, https://flexbooks.ck12.org/cbook/ck-12-precalculus-concepts-2.0/section/16.2/related/lesson/if-then-statements-geom/. To a university audience, Professor Brian Stuart explains: "In other words, the ENIAC is not restricted to control structures that are based on simple counts but allows for logical decisions based on the values of computations." Stuart, "Debugging the ENIAC [Scanning Our Past]," 2340–2341.
13. In her oral history, Kay said that "[t]he master programmer of the ENIAC was the heart and soul of the ENIAC." Antonelli, Oral History, 18. In her ENIAC *Technical Report I, Volume I,* Adele Goldstine writes: "The master programmer is a central programming unit whose primary function is to direct and stimulate the performance of the program sequence of various levels which enter into a computation... It is, however, essential to use the master programmer to accomplish the iteration of a program sequence into a chain (see Section 1.4.) or to link together chains and program sequences." A. Goldstine, ENIAC *Technical Report I, Volume I,* p. X-1.
14. Bartik, Oral History, 44.

Bench Tests and Best Friends

1. While she and Betty were "programming the trajectory," "Marlyn and Ruth were assigned...[the task of] calculating a trajectory one ADD time at a time...so that we would know what was in every accumulator at every ADD time. Well...it took a very long time to do this." Bartik, Oral History, 14.
2. Antonelli, Oral History, 12.
3. Meltzer, Oral History, 22.
4. Stacia Friedman, "Historic Jewish Women's Shelter Transformed Into Lux Apartments," Hidden City, https://hiddencityphila.org/2020/05/historic-jewish-womens-shelter-transformed-into-lux-apartments/.
5. Meltzer, Oral History, 12–13.

6. Meltzer, Oral History, 13.
7. Joy Meltzer, in interview with author, 2021.
8. Meltzer, Oral History, 13.

Parallel Programming

1. Bartik, *Pioneer Programmer*, 84–85.
2. Bartik, Oral History, 11.
3. Bartik, Oral History, 11.
4. Holberton, Oral History, 7.
5. Holberton, Oral History, 7.
6. Holberton, Oral History, 12.
7. Betty asked Nick Metropolis about the lack of accuracy in the calculations they were running for Los Alamos on ENIAC and he responded that, for the purposes of the Los Alamos work, "we just [need to] get in the ballpark." Holberton, Oral History, 4.
8. Kay recalled that when preparing parallel programming for ENIAC, "you had to make sure you didn't start the next part of the program until all of the other things that were running parallel had actually finished.... You had to worry about that kind of thing." Antonelli, Oral History, 16.
9. Antonelli, Oral History, 16.
10. Bartik, *Pioneer Programmer*, 89.
11. "Gale-Battered Ships Arrive with Troops," *Philadelphia Inquirer*, December 25, 1945, 1.

Sines and Cosines

1. Shurkin, *Engines of the Mind*, 197.
2. Henry Herbert, Publicity Manager, University of Pennsylvania, "Demonstration of ENIAC," February 1, 1946, University of Pennsylvania Archives. Herbert notes that he provides this summary of events "[a]t the request of several persons who attended the ENIAC press conference."
3. War Department, Bureau of Public Relations, Press Branch, "Military Applications of ENIAC Described," "High Speed, General Purpose Computing Machines Needed," "Physical Aspects, Operation of ENIAC Are Described," 1946, University of Pennsylvania Archives.
4. In addition, the three men wrote more releases for the War Department, Bureau of Public Relations, "Industrial and Scientific Applications of the ENIAC" and "History of Development of Computing Devices," 1946, University of Pennsylvania Archives as part of their general description of ENIAC and its uses.

5. War Department, "Profiles of Personnel Who Developed ENIAC," 1946, University of Pennsylvania Archives.
6. McCartney, *ENIAC*, 129–130.
7. War Department, "Profiles of Personnel Who Developed ENIAC," 4.
8. The "embargo" warnings vary slightly on each press release, but the overall goal is clear: use of this material by the press is restricted for two weeks from newspaper publication or radio broadcast.
9. Demonstration of ENIAC, February 1, 1946, written by Henry Herbert, publicity manager, University of Pennsylvania, "FOR RELEASE FEBRUARY 16, 1946." Copy in author's collection.
10. Stephen Falk, photographer of ENIAC and ENIAC Programmers, in interview with author, February 1996.
11. University of Pennsylvania Archives, Digital Image Collection, https://library.artstor.org/#/asset/SS7732016_7732016_12329883;prevRouteTS=1641178556553.
12. University of Pennsylvania Archives, Digital Image Collection, https://library.artstor.org/#/asset/SS7732016_7732016_12331236;prevRouteTS=1641178887083.
13. Computer History Museum, https://www.computerhistory.org/collections/catalog/102622392.
14. Teitelbaum, Deposition, 3.
15. Goldstine, *The Computer from Pascal to Von Neumann*, 228.
16. Goldstine, *The Computer from Pascal to Von Neumann*, 228.
17. Teitelbaum, Deposition, 3.
18. Holberton, Oral History, 9.
19. Holberton, Oral History, 9.
20. Antonelli, Oral History, 21.
21. Bartik, *Pioneer Programmer*, 91
22. Bartik, *Pioneer Programmer*, 91.
23. Holberton, Oral History, 54.
24. Bartik, Oral History,16.
25. Smithsonian Oral History of Bartik and Holberton, 1973, 50.
26. Smithsonian Oral History of Bartik and Holberton, 1973, 51.
27. Bartik, Oral History, 16.
28. Bartik, *Pioneer Programmer*, 92.

The ENIAC Room Is Theirs!

1. Holberton, Oral History, 7.
2. Smithsonian Oral History of Bartik and Holberton, 1973, 50.
3. Holberton, Oral History, 6.
4. Bartik, in interview with author, 1996.

5. Brian Stuart, "Programming the ENIAC [Scanning Our Past]," *Proceedings of the IEEE* 106, no. 9 (2018): 1760–70.
6. Holberton, Oral History, 3.
7. Holberton, Oral History, 3.
8. Holberton, Oral History, 7.
9. Holberton, Oral History, 7.
10. For example, "Testing," Bitesize, https://www.bbc.co.uk/bitesize/guides/zg4j7ty/revision/5.
11. Jean recalled, "We came back [to the ENIAC room to work on their program] the next day and we'd find the wires weren't where we thought they were..." Smithsonian Oral History of Bartik and Holberton, 1973, 52.
12. Smithsonian Oral History of Bartik and Holberton, 1973.
13. Meltzer, Oral History, 16.
14. Bartik, Oral History, 15.
15. Bartik, Oral History, 14–15.
16. Bartik, *Pioneer Programmer*, 85.
17. Bartik, *Pioneer Programmer*, 85.
18. Bartik, Oral History, 24.
19. Shurkin, *Engines of the Mind*, 149.
20. Jean Jennings Bartik, "My Personal Impressions." Unpublished notes shared with author.
21. Bartik, *Pioneer Programmer*, 86.
22. Shurkin, *Engines of the Mind*, 154–155.
23. Bartik, *Pioneer Programmer*, 89.

The Last Bugs Before Demonstration Day

1. Holberton, Oral History, 9.
2. Bartik, Oral History, 16.
3. Holberton, Oral History, 9.
4. Bartik, *Pioneer Programmer*, 95.
5. Holberton, Oral History, 9.
6. Bartik, *Pioneer Programmer*, 95.
7. Bartik, *Pioneer Programmer*, 96.
8. Bartik, *Pioneer Programmer*, 96.
9. Bartik, *Pioneer Programmer*, 85.
10. Holberton, Oral History, 9.
11. Holberton, Oral History, 9.
12. Bartik, *Pioneer Programmer*, 96.
13. Antonelli, Oral History, 22.

Demonstration Day, February 15, 1946

1. Holberton, Oral History, 10.
2. See, e.g., Dilys Winegrad and Atsushi Akera, *A Short History of the Second American Revolution*, University of Pennsylvania, 50th anniversary celebration, https://almanac.upenn.edu/archive/v42/n18/eniac.html.
3. Bartik, *Pioneer Programmer*, 98–99.
4. Bartik, Oral History, 17.
5. Antonelli, Oral History, 22.
6. Bartik, Oral History, 15.
7. Bartik, Oral History, 24.
8. Bartik, *Pioneer Programmer*, 96–97. For years, the image of a computer in Hollywood would be one of a large machine with small lights flashing.
9. Bartik, Oral History, 17.
10. Holberton, Oral History, 11.
11. Meltzer, Oral History, 19.
12. Antonelli, Oral History, 22.
13. Antonelli, Oral History, 22.
14. Antonelli, Oral History, 23.
15. Meltzer, Oral History, 19.
16. Bartik, *Pioneer Programmer*, 99.
17. Bartik, *Pioneer Programmer*, 99.
18. Bartik, *Pioneer Programmer*, 99.
19. "U. of P. Exhibits Electronic 'Brain,'" *Philadelphia Inquirer*, February 16, 1946, 3.
20. "U. of P. Exhibits Electronic 'Brain.'"
21. "U. of P. Exhibits Electronic 'Brain.'"
22. Bartik, *Pioneer Programmer*, 99.
23. Menu of February 15, 1946, banquet shown in Erickson, *Top Secret Rosies*.
24. "Blinkin' ENIAC's a Blinkin' Whiz," *Philadelphia Record*, February 15, 1946, 1. From the collection of Marlyn Wescoff Meltzer.
25. Meltzer, Oral History, 19.
26. Holberton, Oral History, 10.
27. Holberton, Oral History, 10.

A Strange Afterparty

1. Meltzer collection of Demonstration Day newspaper clippings shared with author.
2. Holberton, Oral History, 10.
3. "World's Fastest Calculator Cuts Years' Task to Hours," *Boston Globe*, February 15, 1946, 7.

4. "Betty [Jean] Jennings in Scientific Work," *Stanberry Herald-Headlight* (Missouri), March 14, 1946, 1.
5. Holberton, Oral History, p.10.
6. Antonelli, Oral History, 22.
7. This clip can be seen in numerous documentaries, including *The Computers: The Remarkable Story of the ENIAC Programmers*, produced by Kathy Kleiman, Jon Palfreman, and Kate McMahon, 2014.
8. *The Computers: The Remarkable Story of the ENIAC Programmers*.
9. Bartik, *Pioneer Programmer*, 99.
10. Shurkin, *Engines of the Mind*, 198.
11. Shurkin, *Engines of the Mind*, 199.
12. Shurkin, *Engines of the Mind*, 199.
13. Shurkin, *Engines of the Mind*, 199.
14. Shurkin, *Engines of the Mind*, 200.
15. Shurkin, *Engines of the Mind*, 200.
16. Bartik, *Pioneer Programmer*, 101.
17. Shurkin, *Engines of the Mind*, 201. Computer historian Dr. Paul Ceruzzi of the Smithsonian similarly wrote: "The Philadelphia-Princeton region, once a contender for the title of center for computing technology, never recovered." Paul E. Ceruzzi, *A History of Modern Computing* (Cambridge, MA: MIT Press, 1998), 25.

Hundred-Year Problems and Programmers Needed

1. Shurkin, *Engines of the Mind*, 197.
2. Antonelli, Oral History, 27.
3. Antonelli, Oral History, 119.
4. Antonelli, "The KMMA Story."
5. Correspondence of Kathleen McNulty (Mauchly Antonelli) and Douglas Hartree, UPD 8.12, ENIAC Trial Exhibits Master Collection, University Archives & Records Center, University of Pennsylvania.
6. See, e.g., Rand Corporation report on thermodynamic properties of real gases, https://www.rand.org/content/dam/rand/pubs/research_memoranda/2009/RM442.pdf.
7. Holberton, Oral History, 11.
8. Bartik, Oral History, 12.
9. Bartik, Oral History, 12.
10. Bartik, *Pioneer Programmer*, 106.
11. Jean remembered Adele fondly: "I adored Adele. She and I had a very close relationship. She was quite a bit younger than Herman, and we could giggle like young girls together. We had a lot of fun with each other." Bartik, Oral History, 12.

12. Hans Rademacher, "On the Accumulation of Errors in Numerical Integration on the ENIAC," Lecture 19, July 22, 1946, *The Moore School Lectures*, Charles Babbage Institute Reprint Series for the History of Computing, Volume 9, 223–238.
13. D. R. Hartree, "The Eniac, an Electronic Computing Machine," *Nature* 158 (1946): 500–506.
14. "When he got back to England, he discovered a major mistake in his analysis of his problem. Thus, his results were invalid. He wanted changes made to the program and rerun." Antonelli, "The KMMA Story."
15. Hartree, "The Eniac, an Electronic Computing Machine."
16. Antonelli, Oral History, 28. See also correspondence of Kathleen McNulty (Mauchly Antonelli) and Douglas Hartree, UPD 8.12, ENIAC Trial Exhibits Master Collection, University Archives & Records Center, University of Pennsylvania.
17. A. H. Taub, "Refraction of Plane Shock Waves," *Physical Review* 72 (1947): 51–60.
18. In his "A Survey of Eniac Operations and Problems, 1946–1952," W. Barkley Fritz shared information about Dean Goff's work, which he described as "Zero—Pressure Properties of Diatomic Gases—(University of Pennsylvania)." Fritz, "A Survey of Eniac Operations and Problems, 1946–1952," Report No. 617, Ballistic Research Laboratories, Aberdeen Proving Ground, MD, August 1952, 23–24.
19. Fritz, "ENIAC—A Problem Solver," 26.
20. Meltzer, Oral History, 20.

The Moore School Lectures

1. Bartik, *Pioneer Programmer*, 104.
2. Bartik, Oral History, 25.
3. Joy Meltzer, in interview with author, 2020–2021.
4. The twenty-eight students registered for the course can be found in *The Moore School Lectures*, Charles Babbage Institute Reprint Series for the History of Computing, Volume 9, xvi–xvii.
5. *The Moore School Lectures*, ix–xvii.
6. *The Moore School Lectures*, ix–xvii.
7. Holberton, Oral History, 11.
8. *The Moore School Lectures*, ix–xvii.
9. For example, Maurice Wilkes, British computer scientist, https://www.britannica.com/biography/Maurice-Wilkes.
10. McCartney, *ENIAC*, 142.
11. "Wife Drowns in Night Swim with Scientist," *Philadelphia Inquirer*, September 9, 1946, 1.
12. "Wife's Drowning Called Accident," *Philadelphia Inquirer*, September 17, 1946, 3.
13. Bartik, *Pioneer Programmer*, 109.

Their Own Adventures

1. Meltzer, Oral History, 11.
2. Before the Revolution: Socialites and Celebrities Flocked to Cuba in the 1950s, https://www.smithsonianmag.com/history/before-the-revolution-159682020/.
3. Bartik, *Pioneer Programmer*, 106.
4. Bartik, *Pioneer Programmer*, 107.
5. Bartik, *Pioneer Programmer*, 107.
6. Bartik, *Pioneer Programmer*, 130.
7. Antonelli, Oral History, 29.
8. Antonelli, "The KMMA Story."
9. Antonelli, "The KMMA Story."
10. Antonelli, "The KMMA Story."
11. Holberton, Oral History.
12. Antonelli, "The KMMA Story."
13. Antonelli, "The KMMA Story."
14. Antonelli, "The KMMA Story."
15. Antonelli, "The KMMA Story."
16. Antonelli, Oral History, 30.
17. Still Route 66 took them through magnificent locations, including Arizona's Painted Desert and the Grand Canyon. Antonelli, "The KMMA Story."
18. Holberton, Oral History, 12.
19. Holberton, Oral History, 11.
20. Holberton, Oral History, 11.

ENIAC 5 in and around Aberdeen

1. Bartik, *Pioneer Programmer*, 111.
2. Bartik, Oral History, 10.
3. Bartik, *Pioneer Programmer*, 110.
4. Antonelli, Oral History, 29.
5. Holberton, Oral History, 13.
6. Bartik, Oral History, 18.
7. Joy Meltzer, in interview with author, April 29, 2021.
8. Meltzer, Oral History, 20.
9. Teitelbaum, in interview with family, 1986.
10. Robert Sheroke, computer scientist at U.S. Army Research Laboratory, in interview with author, undated.
11. Antonelli, Oral History, 30–31.
12. Antonelli, Oral History, 31.
13. Antonelli, Oral History, 30.
14. Teitelbaum, in interview with family, 1986.

15. Fritz, "The Women of ENIAC," 23.
16. Fritz, "The Women of ENIAC," 15.
17. Fritz, "The Women of ENIAC," 24.
18. Holberton, Oral History, 13.
19. Holberton, Oral History, 13.
20. Bartik, Oral History, 18.
21. Bartik, *Pioneer Programmer*, 115.
22. Bartik, *Pioneer Programmer*, 116.
23. Bartik, Oral History, 18.
24. See, e.g., NASA web page on wind tunnels, https://www.nasa.gov/audience/forstudents/k-4/stories/nasa-knows/what-are-wind-tunnels-k4.html.
25. Bartik, Oral History, 17–18.
26. Clippinger, "A Logical Coding System Applied to the ENIAC" (with Orders for 60 Word Vocabulary, dated November 13, 1947, and test programs included).
27. Bartik, Oral History, 28.
28. Bartik, *Pioneer Programmer*, 116.
29. Bartik, Oral History, 26.
30. Bartik, Oral History, 28.
31. Bartik, *Pioneer Programmer*, 117.
32. Bartik, *Pioneer Programmer*, 117.
33. Bartik, Oral History, 26.
34. Bartik, Oral History, 26.
35. Bartik, *Pioneer Programmer*, 118.
36. Bartik, *Pioneer Programmer*, 118.
37. Bartik, *Pioneer Programmer*, 118–119.
38. Bartik, *Pioneer Programmer*, 119.

A New Life

1. Antonelli, "The KMMA Story."
2. Antonelli, "The KMMA Story."
3. Bartik, *Pioneer Programmer*, 117.
4. W. Barkley Fritz confirms that this division of labor became the accepted breakdown of work at BRL—mathematical, then programming—and was the way teams were comprised at BRL to work on problems for ENIAC. Fritz, "ENIAC—A Problem Solver," 30.
5. Bartik, *Pioneer Programmer*, 117–118.
6. Bartik, *Pioneer Programmer*, 114, 119.
7. Bartik, Oral History, 29.
8. Bartik, *Pioneer Programmer*, 119.

9. Bartik, *Pioneer Programmer*, 120.
10. H. Neukom, "The Second Life of ENIAC," *IEEE Annals of the History of Computing* 28, no. 2 (April–June 2006): 4–16, doi:10.1109/MAHC.2006.39.
11. Thomas Haigh, Mark Priestley, and Crispin Rope, "Engineering 'The Miracle of the ENIAC': Implementing the Modern Code Paradigm," *IEEE Annals of the History of Computing* 36, no. 2 (2014): 47.
12. Antonelli, Oral history, 31.
13. Fritz, "The Women of ENIAC," 24.
14. Fritz, "The Women of ENIAC," 24.
15. Clippinger, "A Logical Coding System Applied to the ENIAC," 1.
16. Patricia Sullivan, "Gloria Gordon Bolotsky, 87," [Obituary piece], *Washington Post*, July 26, 2009.
17. "The Secret History of Women in Coding," *New York Times Magazine*, February 17, 2019 (cover).
18. Fritz, "The Women of ENIAC," 23. Others, including Lila Todd (Butler), would take their maternity leave and return to work at BRL. "The Women of ENIAC," 15.
19. Holberton, Oral History, 14.
20. Bartik, *Pioneer Programmer*, 119.
21. Teitelbaum, in interview with family, 1986.
22. Antonelli, Oral History, 32.
23. Antonelli, Oral History, 32.
24. Antonelli, "The KMMA Story."
25. Fritz, "ENIAC—A Problem Solver," 39.
26. Fritz, "A Survey of Eniac Operations and Problems, 1946–1952."
27. Fritz published comprehensive lists of the many complex, nonclassified problems run on ENIAC by military groups, government agencies, universities, companies, and institutions, including in 1952, "A Survey of Eniac Operations and Problems, 1946–1952," and in 1994, "ENIAC—A Problem Solver," Appendix 41–45.
28. Fritz, "ENIAC—A Problem Solver," 44.
29. Herman's book with a footnote showing a letter from his wife, who was working in Los Alamos. Goldstine, *The Computer from Pascal to Von Neumann*.
30. Fritz, "ENIAC—A Problem Solver," 40.
31. Fritz, "ENIAC—A Problem Solver," 33.

Epilogue

1. Shurkin, *Engines of the Mind*, footnote 13, 230–231.
2. Nathan Ensmenger, *The Computer Boys Take Over* (Cambridge, MA: MIT Press, 2010): 35–36.
3. Cory Doctorow, "The ENIAC Programmers: How Women Invented Modern

Programming and Then Were Written Out of the History Books," *Boing Boing*, June 21, 2019, https://boingboing.net/2019/06/21/founding-mothers-of-computing.html.

4. "Black Women Oral History Project Interviews, 1976–1981," Harvard Radcliffe Institute, https://guides.library.harvard.edu/schlesinger_bwohp.

Postscript

1. Antonelli, Oral History, 33.
2. Antonelli, "The KMMA Story."
3. Gini Mauchly Calcerano, in interview with author, 2018.
4. Antonelli, "The KMMA Story."
5. There are many other honors, and Gini Mauchly Calcerano would have a full and up-to-date list.
6. Jay and David Teitelbaum, in interview with author and Amy Sohn, May 25, 2020.
7. Through Kay, the author learned that Fran's husband, Homer Spence, was ill in and around the fiftieth anniversary of ENIAC in 1996. After Fran passed away in 2012, her sister Judith Veitch shared some family history in a letter to the author. Her sons and daughters-in-law also shared photographs of Fran as a girl, in college, and with her husband and sons.
8. Meltzer, Oral History, 33.
9. Jean Bartik, "Minicomputers Turn Classic," Auerbach Scientific and Control Computers Reports, *Data Processing Magazine*, January 1970, 42.
10. Jean Bartik's "10 Proverbs of Life," shared with the author and delivered as part of her commencement address to Northwest Missouri State University on April 27, 2002.
11. Holberton, Oral History, 18.
12. "ENIAC Hall Of Fame Inductees," 1997, WITI, video, https://www.youtube.com/watch?v=DsctkUrUYgo.

Selected Bibliography

Books

Bartik, Jean Jennings. *Pioneer Programmer: Jean Jennings Bartik and the Computer that Changed the World*. Kirksville, MO: Truman State University Press, 2013.

Bidwell, Shelford, ed. *Brassey's Artillery of the World: Guns, Howitzers, Mortars, Guided Weapons, Rockets and Ancillary Equipment in Service with the Regular and Reserve Forces of All Nations*. London: Brassey's Publishers Ltd., 1977.

Campbell-Kelly, Martin, and Michael R. Williams, eds. *The Moore School Lectures: Theory and Techniques for Design of Electronic Digital Computers*, volume 9 in the Charles Babbage Institute Reprint Series for the History of Computing. Cambridge, MA: MIT Press, 1985.

Ceruzzi, Paul E. *A History of Modern Computing*. Cambridge, MA: MIT Press, 1998.

Goldstine, Herman H. *The Computer from Pascal to Von Neumann*. Princeton, NJ: Princeton University Press, 1993.

Grier, David Alan. *When Computers Were Human*. Princeton, NJ: Princeton University Press, 2005.

Isaacson, Walter. *The Innovators: How a Group of Hackers, Geniuses, and Geeks Created the Digital Revolution*. New York: Simon & Schuster, 2014.

Knuth, D. E. *The Art of Computer Programming*, Vol. 3, *Sorting and Searching*. London: Pearson Education, 1998.

Lee, J. A. N. "John Grist Brainerd" in *Computer Pioneers*. Los Alamitos, CA: IEEE Computer Society Press, 1995.

McCartney, Scott. *ENIAC, the Triumphs and Tragedies of the World's First Computer*. New York: Walker and Company, 1999.

Metropolis, N., J. Howlett, and Gian-Carlo Rota, eds. *A History of Computing in the Twentieth Century: A Collection of Essays*. New York: Academic Press, 1980.

Rhodes, Richard. *Dark Sun: The Making of the Hydrogen Bomb*. New York: Simon & Schuster, 1995.

Rossiter, Margaret W. *Women Scientists in America, Before Affirmative Action, 1940–1972*. Baltimore: Johns Hopkins University Press, 1998.

———. *Women Scientists in America, Struggles and Strategies to 1940*. Baltimore: Johns Hopkins University Press, 1982.

Shurkin, J. N. *Engines of the Mind: The Evolution of the Computer from Mainframes to Microprocessors*. New York: W. W. Norton & Company, 1996.

Oral Histories

Antonelli, Kathleen "Kay" McNulty Mauchly. Interview by author and directed by David Roland. Recorded in the home of Mrs. Antonelli, September 18, 1997. Transcript. ENIAC Programmers Oral History Project.

Bartik, Jean Jennings. Interview by author and directed by David Roland. Recorded in the home of Ms. Bartik, September 17, 1997. Transcript. ENIAC Programmers Oral History Project.

Bartik, Jean J., and Frances E. (Betty) Snyder Holberton. Interview by Henry S. Tropp. Transcript. Computer Oral History Collection, 1969–1973, 1977. Archives Center, Smithsonian National Museum of American History, April 27, 1973. https://amhistory.si.edu/archives/AC0196_bart730427.pdf.

Burks, Alice R., and Arthur W. Burks. Interview by Nancy Stern. Oral history interview with Alice R. Burks and Arthur W. Burks. Charles Babbage Institute, Oral History, Center for the History of Information Processing, University of Minnesota, Minneapolis, June 20, 1980. https://conservancy.umn.edu/handle/11299/107206.

Eckert, J. Presper. Interview by Nancy Stern. Oral history interview with J. Presper Eckert. Sperry Univac (Blue Bell, PA), Charles Babbage Institute, Center for the History of Information Processing, University of Minnesota, Minneapolis, October 28, 1977. https://conservancy.umn.edu/handle/11299/107275.

Holberton, Frances Elizabeth "Betty" Snyder. Interview by author and directed by David Roland. Recorded in the library of the Shady Grove Center, Rockville, MD, September 23–24, 1997. Transcript. ENIAC Programmers Oral History Project.

Mauchly, John. Autobiographical interview, part 3. Interview by Esther Carr. Video. 1977. https://www.youtube.com/playlist?list=PL0IDvwajM_78cEx-KaJdixj8cFFuC8FUC.

———. Interviewed by Nancy Stern. Video. Niels Bohn Library and Archives Oral Histories, American Institute of Physics, Friday, May 6, 1977. https://www.aip.org/history-programs/niels-bohr-library/oral-histories/31773.

Meltzer, Marlyn Wescoff. Interview by author and directed by David Roland. Recorded in the home of Mrs. Meltzer, September 16, 1997. Transcript. ENIAC Programmers Oral History Project.

Teitelbaum, Ruth. Interview by Adolph, Jay, and David Teitelbaum. Recorded in Teitelbaum home, July 18, 1986. Oral History of Ruth Lichterman Teitelbaum.

Selected Interviews

Antonelli, Kathleen "Kay" McNulty Mauchly. Interview with author, April 18, 2000.

———. Interview with author, July 20, 2003.

Atwater, Dr. William. Director, with author, at U.S. Army Ordnance Museum, Aberdeen, MD, undated.

Benson, Josephine. Interview with author and Amy Sohn via Zoom, February 29, 2020.

Calcerano, Gini Mauchly. Interview with author and Amy Sohn via Zoom, May 25, 2021.

Falk, Steven M. Interview with author, Philadelphia, February 14, 1996.

Madlen Simon. Daughter of Adele and Herman Goldstine. Interview with author and Amy Sohn via Zoom, June 15, 2020.

Meltzer, Joy. Interview with author via phone, April 29, 2021.

Meltzer, Marlyn. Interview with author, February 6, 1996.

———. Interview with Thomas Petzinger Jr., 1996.

Teitelbaum, Jay, David Teitelbaum, Melinda Teitelbaum, and Suzanne Teitelbaum. Interview with author and Amy Sohn via Zoom, May 28, 2021.

Teitelbaum, Jay, and David Teitelbaum. Interview with author and Amy Sohn via Skype video, May 25, 2020.

Articles (Published Works)

Ceruzzi, Paul E. "When Computers Were Human." *Annals of the History of Computing* 13, no. 3 (1991): 237–44.

Costello, J. "As the Twig Is Bent: The Early Life of John Mauchly." *IEEE Annals of the History of Computing* 18, no. 1 (1996): 45–50.

Eckstein, P. "J. Presper Eckert." *IEEE Annals of the History of Computing* 18, no. 1 (1996): 25–44.

Fritz, W. Barkley. "ENIAC—A Problem Solver." *IEEE Annals of the History of Computing* 16, no. 1 (March 1994): 25–45.

———. "The Women of ENIAC." *IEEE Annals of the History of Computing* 18, no. 3 (1996): 13–28.

Grier, David. "The ENIAC, the Verb 'to Program' and the Emergence of Digital Computers." *IEEE Annals of the History of Computing* 18, no. 1 (March 1996): 51–55.

Haigh, Thomas, Mark Priestley, and Crispin Rope. "Engineering 'The Miracle of the ENIAC': Implementing the Modern Code Paradigm." *IEEE Annals of the History of Computing* 36, no. 2 (2014): 41–59.

Hartree, D. R. "The Eniac, an Electronic Computing Machine." *Nature* 158, no. 4015 (October 1, 1946): 500–506.

Light, Jennifer S. "When Computers Were Women." *Technology and Culture* 40, no. 3 (1999): 455–83.

Mauchly, John W. "Amending the ENIAC Story." *Datamation* 25, no. 11 (1979).

———. "Mauchly: Unpublished Remarks." *IEEE Annals of the History of Computing* 4, no. 3 (July 1982): 245–56.

Mauchly, Kathleen R. "John Mauchly's Early Years." *IEEE Annals of the History of Computing* 6, no. 2 (1984): 116–38.

Metropolis, N., and E. C. Nelson. "Early Computing at Los Alamos." *IEEE Annals of the History of Computing* 4, no. 4 (1982): 348–57.

Metropolis, N., and J. Worlton. "A Trilogy on Errors in the History of Computing." *IEEE Annals of the History of Computing* 2, no. 1 (1980): 49–59.

Neukom, H. "The Second Life of ENIAC." *IEEE Annals of the History of Computing* 28, no. 2 (April 2006): 4–16.

Seabright, McCabe. "Adele Goldstine: The Woman Who Wrote the Book." *SWE Magazine*, Spring 2019.

———. "Finding Alyce Hall." *SWE Magazine*, 2014.

Stern, Nancy. "John William Mauchly, 1907–1980." *IEEE Annals of the History of Computing* 2, no. 2 (1980): 100–103.

Stuart, Brian L. "Debugging the ENIAC [Scanning Our Past]." *Proceedings of the IEEE* 106, no. 12 (2018): 2331–45.

———. "Programming the ENIAC [Scanning Our Past]." *Proceedings of the IEEE* 106, no. 9 (2018): 1760–70.

———. "Simulating the ENIAC [Scanning Our Past]." *Proceedings of the IEEE* 106, no. 4 (2018): 761–72.

Taub, A. H. "Refraction of Plane Shock Waves." *Physical Review* 72, no. 1 (July 1, 1947): 51–60.

Weik, Martin H. "The ENIAC Story." *Ordnance* 45, no. 244 (1961): 571–75.

Winegrad, Dilys. "Celebrating the Birth of Modern Computing: The Fiftieth Anniversary of a Discovery at the Moore School of Engineering of the University of Pennsylvania." *IEEE Annals of the History of Computing* 18, no. 1 (March 1996): 5–9.

Articles, Pamphlets, Essays, and Speeches (Unpublished Works)

"Deposition of Ruth Teitelbaum," ENIAC Patent Trial Collection, Box 16, University of Pennsylvania Archives.

Antonelli, Kathleen McNulty Mauchly. "Compressible Boundary Layer, Zero-order functions. Set-up for programming of integrations." Pedaling Sheets. 1946.

———. "The Kathleen McNulty Mauchly Antonelli Story." March 26, 2004. https://sites.google.com/a/opgate.com/eniac/Home/kay-mcnulty-mauchly-antonelli.

———. "Luncheon Speech, Reminiscences." Introduction Speech at the 40th Anniversary of ENIAC. Transcript by author. October 1986.

Bergin, Thomas J., ed. *50 Years of Army Computing, From ENIAC to MSRC: A Record of a Symposium and Celebration, November 13 and 14, 1996*. Sponsored by the Army Research Laboratory and U.S. Army Ordnance Center & School, September 2000.

Clippinger, R. F. "A Logical Coding System Applied to the ENIAC," Report No. 673. Ballistic Research Laboratories, Aberdeen, MD, September 29, 1948. https://apps.dtic.mil/sti/citations/ADB205179.

Fritz, W. Barkley. "A Survey of Eniac Operations and Problems, 1946–1952," Report No. 617. Ballistic Research Laboratories, Aberdeen Proving Ground, MD, August 1952. https://apps.dtic.mil/sti/pdfs/AD1003735.pdf.

Goldstine, Adele K., "Report on THE ENIAC (Electronic Numerical Integrator and Computer)." *Technical Report* I (June 1, 1946). Developed under the supervision of the Ordnance Department, United States Army, University of Pennsylvania, Moore School of Electrical Engineering, Philadelphia.

Kleiman, Kathryn. "Biography of Mrs. Frances Elizabeth Snyder Holberton," Fletcher, Heald & Hildreth, Arlington, VA, 1996.

Reid, Henry. "Ballisticians in War and Peace, Volume I, 1914–1956." Army Research Labs, Aberdeen Proving Ground, MD. https://apps.dtic.mil/sti/pdfs/ADA300523.pdf.

Manuscript and Archival Sources

Ancestry.com.

Archives Center, Smithsonian National Museum of American History.

Arthur and Elizabeth Schlesinger Library on the History of Women in America, Radcliffe Institute for Advanced Study, Harvard University.

Charles Babbage Institute, Oral History, Center for the History of Information Processing, University of Minnesota, Minneapolis.

Drexel University Archives.

Jean Jennings Bartik Computing Museum, Northwest Missouri State University.

Kislak Center for Special Collections, Rare Books and Manuscripts, University of Pennsylvania.

Library of Congress, Science, Technology & Business Division.

The National Archives.

Niels Bohn Library and Archives Oral Histories, American Institute of Physics. https://www.aip.org/history-programs/niels-bohr-library/oral-histories/31773.

Social Science & History/Newspaper Department, Free Library of Philadelphia.

Special Collections Research Center, Temple University Libraries Special Collections Research Center.

University Archives and Records Center, University of Pennsylvania.
———. Class Records and Yearbooks.
———. Digital Image Collection.
———. ENIAC Patent Trial Collection.
———. University Relations Information Files.
U.S. Army Ordnance Museum.

Selected Newspaper Articles

Lohr, Steve. "Frances E. Holberton, 84, Early Computer Programmer." *New York Times*, December 17, 2001. https://www.nytimes.com/2001/12/17/business/frances-e-holberton-84-early-computer-programmer.html.

———. "Jean Bartik, Software Pioneer, Dies 86." *New York Times*, April 8, 2011. https://www.nytimes.com/2011/04/08/business/08bartik.html.

Petzinger, Thomas Jr. "Female Pioneers Fostered Practicality of Computers." *Wall Street Journal*, November 22, 1996. https://www.wsj.com/articles/SB848618358629375500.

———. "History of Software Begins with the Work of Some Brainy Women." *Wall Street Journal*, November 15, 1996. https://www.wsj.com/articles/SB848012407846877000.

Saxon, Wolfgang. "Herman Goldstine Dies at 90; Helped Build First Computers." *New York Times*, June 26, 2004. https://www.nytimes.com/2004/06/26/us/herman-goldstine-dies-at-90-helped-build-first-computers.html.

Sullivan, Patricia. "Gloria Gordon Bolotsky, 87; Programmer Worked on Historic ENIAC Computer." *Washington Post*, July 26, 2009. https://www.washingtonpost.com/wp-dyn/content/article/2009/07/25/AR2009072502045.html.

Websites, Movies, and Documentaries

Coughlin, Bill. "Commercial Digital Computer Birthplace." Historical Marker Database. March 14, 2011. Updated June 16, 2016. https://www.hmdb.org/m.asp?m=40918.

David, Paul. *Mauchly: The Computer and the Skateboard*. Video. Documentary. Blast-off media, 2001.

Donohoe, Victoria. "Narberth—A History." Friends of Narberth History. October 14, 1994. https://narberthhistory.org/stories/narberth-history.

Erickson, LeAnne. *Top Secret Rosies*. Video. Documentary. PBS Distribution, 2010. http://topsecretrosies.com/.

Evans, Shawn. "Historic Movie Theaters of Center City." *The PhillyHistory Blog*. February 9, 2011. https://blog.phillyhistory.org/index.php/2011/02/historic-movie-theaters-of-center-city/.

"Farm Journal Magazine." AG Web. https://www.agweb.com/farm-journal-magazine.

Friedman, Stacia. "Historic Jewish Women's Shelter Transformed into Lux Apartments." Hidden City Phila. May 16, 2020. https://hiddencityphila.org/2020/05/historic-jewish-womens-shelter-transformed-into-lux-apartments/.

"History of Chestnut Hill College." Chestnut Hill College. https://www.chc.edu/history-chestnut-hill-college.

"History of ESE at Penn." University of Pennsylvania Department of Electrical and Systems Engineering. https://www.ese.upenn.edu/history/.

"Homes of Families with Members Born in Italy, West Philadelphia, 1940." West Philadelphia Collaborative History. https://collaborativehistory.gse.upenn.edu/media/houses-family-members-born-italy-west-philadelphia-1940.

Kleiman, Kathy. *The Computers: The Remarkable Story of the ENIAC Programmers*. Produced by Kathy Kleiman, Jon Palfreman, and Kate McMahon. Video. Documentary. Women Make Movies distributor, 2014. http://eniacprogrammers.org/see-the-film/.

Lang, Walter. *Desk Set*. Movie. 20th Century Fox, 1957.

Muuss, Michael, ed. "History of Computing Information." The U.S. Army Research Labs. https://ftp.arl.army.mil/~mike/comphist/.

———. "Historic Computer Images," The U.S. Army Research Labs. https://ftp.arl.army.mil/ftp/historic-computers/.

"Northwest History." Northwest Missouri Teachers College. https://www.nwmissouri.edu/aboutus/history.htm.

"Our History." Central High School. School District of Philadelphia. Modified February 19, 2020. https://centralhs.philasd.org/about-central-high-school/about-us/.

"Pine Camp, now Fort Drum, in the 1930s and 40s." A North Country Public Radio Project. https://www.northcountryatwork.org/collections/pine-camp-now-fort-drum-in-the-1930s-and-40s/.

Sklaroff, Susan. "The Rebecca Gratz Club." Rebecca Gratz & 19th-Century America. August 24, 2010. http://rebeccagratz.blogspot.com/2010/08/rebecca-gratz-club.html.

Spring, Kelly A. "In the Military during World War II." National Women's History Museum. 2017. https://www.womenshistory.org/resources/general/military.

Strasser, Mike. "Fort Drum exhibit to highlight sonic deception training at Pine Camp during WWII." U.S. Army, November 2, 2020. https://www.army.mil/article/240486/fort_drum_exhibit_to_highlight_sonic_deception_training_at_pine_camp_during_wwii.

Women in Technology International. ENIAC Keynote @ WITI New York Network Meeting. Six-part video. February 23, 1998. https://www.youtube.com/watch?v=P2AjiPhtoJ0&t=13s.

———. 50th Anniversary induction of ENIAC Programmers into Hall of Fame. Video.

1997. https://www.youtube.com/watch?v=kstqypCpHx8&list=PL9zninK8B_FTo-H_H6BhspjQ8J0QXXu7r.
"Women & World War II." Metropolitan State University of Denver. https://temp.msudenver.edu/camphale/thewomensarmycorps/womenwwii/.
"YWCA in Philadelphia." Temple Digital Collections, Temple University Libraries. https://digital.library.temple.edu/digital/custom/ywcaphiladelphia.

Acknowledgments

It takes a village to tell a military, mathematical, technical, and women's story. The people who have helped me have been far-flung and diverse and constant only in their desire to see this untold story shared.

My thanks to Grand Central Publishing and my editor, Suzanne O'Neill. Thank you for your careful editing and continuing reminder to "keep it in the eyes of the women." Thanks to my agent Richard Abate of 3 Arts Entertainment for excellent guidance.

My appreciation to outstanding college professors Sonya Michel at Harvard University and Joseph Weizenbaum of MIT for providing support and encouragement in the junior and senior papers that launched me on this journey, and Professor Stuart Kurtz at the University of Chicago for teaching tremendous computer science courses.

I appreciate the encouragement of my academic community, including fellows, students, and scholars at Princeton's Center for Information Technology Policy and faculty and friends at American University Washington College of Law's Clinical Program, including Vicki Phillips in the Glushko-Samuelson Intellectual Property Law Clinic and Christine Farley, Michael Carroll, Sean Flynn, and Tahniat Saulat in the Program for Intellectual Property and Information Justice.

Old friends stepped in as expert guides along the way, including Professor David Cruz-Uribe, Chair, University of Alabama Department of Mathematics, who spent hours with me reviewing the ballistics trajectory equations and numerical analysis techniques. Dr. Mitchell Lazarus, electrical engineer, mathematics professor, and telecommunications attorney, who dove into ENIAC diagrams and descriptions and shared his personal experiences with circuit boards and plugboards; his novel, *The Implosion Method*, should be on everyone's WWII reading list. Thanks to you both!

The foundation of this book is the ENIAC Programmers' oral histories, and their research and recording would not have been possible without the support of modern computer pioneers. They include Dr. Barbara Simons, the first woman to graduate with a PhD in computer science from the University of California, Berkeley, and second female president of the Association for Computing Machinery (ACM), who encouraged my history work and ACM sponsorship. Mitch Kapor, founder of Lotus Development Corporation, provided the grant through the Kapor Family Foundation that enabled me to spend months at the Library of Congress Science and Technology Reading Room researching 1940s and 1950s computing technology, and then to assemble a first-class team to tape extensive broadcast-quality oral histories with Marlyn, Kay, Jean, and Betty.

My gratitude to senior PBS producer and past Peabody Awards Chair David Roland for directing the oral history productions and assembling an outstanding production team: Sheila Smith, director of photography; and Mary Keigler, sound engineer. To speak to a crew of mostly women helped put the ENIAC Programmers at ease as they talked about some challenges of their youth.

A decade later, Megan Smith, then VP at Google X and soon to be US Chief Technology Officer, introduced me to Anne Wojcicki and

Lucy Southworth Page, who supported my documentary production and research for this book through their foundations. Thanks to Dr. Jon Palfreman, a veteran of BBC, WGBH, and *Frontline* documentaries, for producing our award-winning documentary, *The Computers: The Remarkable Story of the ENIAC Programmers*, with director Kate McMahon and Mark Rublee, director of photography and editor extraordinaire. Great thanks to our advisers Dr. Telle Whitney, Dr. Tracy Camp, and Dr. Paul Ceruzzi for their guidance and advice.

My thanks would not be complete without acknowledging my many trips to the Army Research Laboratory (ARL) at Aberdeen Proving Ground. Heir to Ballistic Research Laboratory and one of the main Army supercomputing centers today, ARL at APG proudly traces its lineage directly to ENIAC. I appreciated my visits with Charles Nietubitz, Robert Sheroke, and Brian Simmonds. Thank you for honoring Jean in her visits to APG and for helping me explore BRL and APG as they would have operated during WWII. Your naming of six new supercomputers after the ENIAC 6 is a tremendous way of sharing their history.

I'll miss the Army Ordnance Museum, formerly located at Aberdeen Proving Ground, and its longtime director, the indomitable Dr. William Atwater. It was a highlight of my study of WWII artillery to spend a day with you learning about the special team-oriented approach to Army artillery operation during and after WWII, and how revolutions of computers and communications have made US artillery so successful.

Great appreciation to Dr. Brian Stuart, professor of computer science at Drexel University, for his excellent papers "Programming the ENIAC" and "Debugging the ENIAC," both in *Proceedings of the IEEE*, our many discussions, and your generous sharing of historical documents, photographs, advice, and edits.

Thank you to the archivists at the University of Pennsylvania Archives who helped me find the reports, letters, and photographs that helped bring my research to life. Former archivist Gail Pietrzyk dove deep with me into old and unlabeled ENIAC photographs, finding ones that the ENIAC Programmers had not seen in fifty years and helped us to label. More recently, archivist Tim Horning and Jim Duffin provided research and photographs for this book, even during COVID lockdown. Before that they organized a special exhibit for the children of the ENIAC Programmers, showing them letters, photographs, and reports of their mothers, and for some of their fathers, that they had never seen before. It was a special moment.

Thanks also to Lorraine Coons at Chestnut Hill College; Eric Dillalogue at Kislak Center, University of Pennsylvania; Marija Gudauskas, Free Library of Philadelphia; and Josué Hurtado and Margery Sly in Special Collections Research Center, Temple University Libraries.

I appreciate my friendships with later programming pioneers who shared stories about their own work and that of the ENIAC Programmers who came before them. They include the indefatigable Jean Sammet, first female president of ACM and longtime IBM pathmaker, and Milly Koss, vice president of Harvard's Information Technology group and programmer at Eckert-Mauchly Computer Corporation decades earlier at the start of her long computing career.

I hope my appreciation and respect for the ENIAC 6 is written into every sentence of this book. I thank the ENIAC Programmers' children for adopting me and this project. Special thanks to Priscilla Holberton; Joy Meltzer; Gini Mauchly Calcerano; Bill Mauchly; Tim, Jane, and Mary Bartik; David and Jay Teitelbaum; and Joe Spence for answering innumerable questions, finding dates, and digging out old photographs from frames and photo albums. I admire how you support the legacies of your mothers, and fathers, including Gini's trips to Ireland for Kay's awards and Tim's editing and promotion of his mother's autobiography, *Pioneer Programmer: Jean Jennings Bartik and the Computer that Changed the World*. Priscilla, thank you for designing the first website for the ENIAC Programmers Project.

It takes great guidance and support to write a proposal and edit a book, and Tina Cassidy and Stuart Horwitz provided invaluable advice and help along the way. It takes great attorneys to navigate difficult questions, and James Gregorio, Zick Rubin, and Marc Miller shared extraordinary advice and counsel.

Thanks to Helen Edersheim for a close review of this manuscript; Jacqueline Young, Grand Central Publishing; and Martha Stevens, 3 Arts, for answering innumerable questions; and to Annalee Greusel for endnote expertise.

Finally, my family has been an invaluable source of support. My father always asked about the ENIAC Programmers and about my research. From his dissertation work interviewing the founders of the semiconductor industry, interviews now housed in Stanford University Libraries, he knew that oral histories of technology pioneers are valuable to future generations.

My mother shared her perspective that there are many distinct and different perspectives of history, and each teller has a different view. Thank you for your many contributions to this work.

Lieutenant Colonel Jerome Massey, my father-in-law, told stories of WWII and how as a private he fought back and forth across the deserts of North Africa and then marched up the boot of Italy, sailed into France, and then marched into Germany. His stories formed a backdrop to this work.

Special smiles to my children, Sam and Robin, who listened to countless stories of the ENIAC Programmers and created a few of their own. I'll never forget Sam sitting with Jean, their heads together, and laughing as Jean shared patent diagrams of ENIAC units and told funny stories of ENIAC work. On a different date, Robin skipped up to Jean with a small Nintendo DS cradled in his hands and laughed as Jean shared stories of the huge size of ENIAC. Later Robin became a dedicated reviewer of the documentary and book, and Sam shares stories of the ENIAC Programmers with his computer science students. Thank you to my brother Steve for endless advice and technical support, and to my stepson Aaron for carrying countless boxes of photos and files to and from my writing office.

My biggest appreciation goes to my husband and partner, Mark Massey. A self-described "old techie," you joined me on this journey fourteen years ago and became my sounding board, supporter, encourager, fact-checker, technology debugger, editor, and discussant in all aspects of this story.

Lastly, thank you to all who contributed to the research and telling of this story and who use the story of the ENIAC 6 to tell your own stories, share your challenges, build your networks, and find support for your work in the computing field. Please continue to tell stories of grandmothers and great-aunts who inspired you to enter computing, and to encourage your daughters and sons to explore careers in science, technology, engineering, and mathematics (STEM). It takes a global village to tell an untold story of programming pioneers and to open doors to STEM careers for everyone.

Index

Note: Italic page numbers refer to illustrations.

About the Author

Kathy Kleiman is an attorney and professor of Internet policy and intellectual property—and coproducer of the award-winning documentary *The Computers: The Remarkable Story of the ENIAC Programmers.*

Kleiman helped found the Internet Corporation for Assigned Names and Numbers (ICANN), the organization that manages and oversees the Internet, and is cofounder of ICANN's Noncommercial Users Constituency. Never afraid of a challenge, she spends her time leading thought-provoking seminars, tracking down additional information about the ENIAC Programmers and sharing their stories with audiences worldwide, and advocating for free speech, fair use, and privacy in global Internet policies.

She teaches Internet technology and governance at American University Washington College of Law and serves as a Faculty Fellow of AU's Internet Governance Lab.

For her work uncovering and preserving the ENIAC Programmers' story, Kleiman has been recognized by the US Army Research Laboratory, and the March of Dimes named her a Lifetime Heroine in Technology.

About the Author

[illegible]

[illegible]

[illegible]